Contents

Introduction

A very unscientific introduction

Hello! *Dogology* is a celebration of the 0.5–1 billion* furry, stinky-breathed, wet-nosed, face-licking, shoe-chewing, manipulative, gorgeous four-legged mongrels, pedigrees, poo machines and assorted fluffy woo-woos that have our hearts wrapped around their little dewclaws. It's also a celebration of the bizarre, fascinating and often hilarious science behind your own domestic zoology project, who has a breathtaking appetite for play, for tummy-tickling and for sharing the rush of hormones we call love, despite being 99.96% wolf.

As a kid I was always desperate, desperate, DESPERATE for a lovely, scruffy, tail-wagging dog. So my parents bought me a gerbil. If you've never had a gerbil, they are basically a poor man's hamster (which is, of course, a poor man's guinea pig, which is a poor man's rabbit, which is a poor man's cat, which is a poor man's dog**). Gerbils are much less cuddly than hamsters and much more rat-like, which my parents convinced me was a good thing. I was actually pretty chuffed with Gerald the Gerbil, though, because I never dared imagine that we could afford a dog, but I constantly dreamed of what might be. Furriness was less of a draw than the idea of

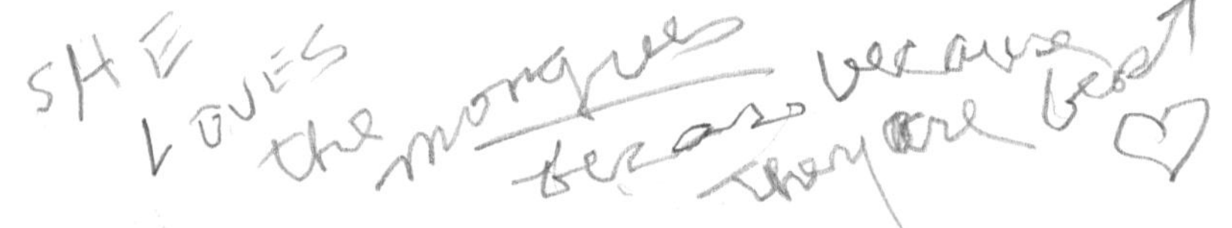

companionship. I was a small-town boy drowning in boredom and a dog would not only love me unconditionally, but more importantly it would be a partner. I *knew* that if I had a dog we'd spend our days rescuing injured ramblers, unearthing treasure, solving crime, helping old folk and putting out fires, before ending the day, exhausted but happy, overlooking the canyon together at sunset. You simply don't get that with a gerbil. Not in Milton Keynes, anyway.

It's traditional for writers of books about dogs to bang on about how wonderful their own dog is, but the sheer volume of facts packed into this book is already giving my publisher paper-budget hives, so I'll be brief. Now that I'm a card-carrying grown-up, I have a dog. My gorgeous scruffy mutt Blue is a Border Collie-Poodle cross and I love him to pieces. Weirdly, he's as obsessed with balls as he is disinterested in food (with the exception of his morning croissant). We don't get up to a lot of canyon-gazing, but our adventures are plentiful: we explore, we meander, we play. He's also a sheer sensory onslaught: unbearably cuddlable, beautiful, and he loves me unconditionally with those big adoring eyes, as requested.

But the best thing about Blue is that he makes me a better person. Not in any quantifiable, scientific way – I'm just nicer, more thoughtful, and care more about my family, my friends, my world and the people in it. We sometimes forget how much of a privilege it

*How many dogs are there? Simple question. Hellish to answer. Statistics on the global domestic dog population vary and use wildly different methodology. A decent estimate is 0.5–1 billion.

** I don't mean that cats are any less important than dogs – they are literally cheaper to run. See p112.

is to share our homes with a large predatory mammal. I've got three other lovely mammals in the house already – one's 16, the other's 18, and the third doesn't like me to be too specific – but Blue is a different species, and it's rare for different species to live so closely together. (There's also a cat, but she's the subject of my other book, *Catology*.) Dogs have, in evolutionary terms, only recently wandered into our homes for warmth, love and regular meals, and they aren't far removed from wild, vicious, ungulate-munching pack animals that would tear your face off as soon as they look at you. (Actually, wolves are much more complex than that, but you get my point.)

Sharing our lives with an entirely different species helps us realize what it is to be human. We make a fundamental shift in communication, expectations, patience, voice, emotions and sense of right and wrong when we relate to dogs. They make us aware of our own extraordinary powers of abstract thought, our desire to nurture, our compassion and empathy, and our great power and responsibility – our ability to transform the world, change the climate and affect the other species we share this place with. Dogs remind us that we need to walk lightly on this Earth.

Thank you so much for reading this book. I'm a member of an odd but lovely little gang of people called science communicators, and we get a huge amount of pleasure not just from telling you amazing things, but from making learning exhilarating. You'll find us at science festivals, comedy clubs, in schools, on the telly, in pubs and in the kitchen at parties. If there's one thing we'd like you to take away from all this knowledge, it's that science can be fascinating, shocking, revelatory and often very, very funny. If you spot one of us on the street, do come and say hi but beware: we are avid collectors of facts and we've got so much to tell you.

Note

There are obviously many different canine subspecies including foxes, dingoes and African wild dogs. For brevity, whenever the word 'dog' is used, I'm talking about the domestic dog (*Canis familiaris*), unless I mention otherwise.

Disclaimer

Nothing in this book is supposed to represent veterinary advice, behavioural advice or training advice. If you have any concerns about your dog, please visit a registered vet or animal behaviourist.

Please ...

Be kind to animals and remember that their experience of the world and their perception is very different to ours. And always pick up your dog's poo. If there's anything more likely to make people hate our dogs, it's stepping in a hot, wet pile of shit.

Chapter 02: What Is a Dog?

2.01 A brief history of the dog

Many facts, dates and places relating to dogs' evolution and domestication are hotly disputed. We do know that in evolutionary terms, dogs are relatively young at 20,000–40,000 years old, and that they're descended from wolves, which first appeared in North America 300,000 years ago (around the same time as humans appeared in Africa). The dog's closest living relative is the grey wolf, *Canis lupus*, but this is a sister group and dogs' immediate ancestors are unknown and probably extinct. Most dog breeds only developed 150–200 years ago.

65 MYA

Dinosaurs die out at the end of the Cretaceous period after a successful 165 million years

55 MYA

Carnivorous mammals emerge

50 MYA

Carnivores diverge into wolflike caniforms and catlike feliforms

3–1 MYA

Wolflike members of the *Canis* genus evolve in Eurasia

300,000 BCE

Wolves appear in North America

Homo sapiens appears in Africa

40,000–20,000 BCE

Modern dogs begin to diverge from wolves

23,000 BCE

Possible date of dog domestication in Siberia

15,000 BCE

Dogs now fully diverged from wolves

14,223 BCE

Age of oldest evidence of pet dog ownership

12,000–10,000 BCE

Dog body size decreases by 38–46% (probably due to domestication)

11,000 BCE

Clear evidence of human/canine cohabitations

9,500 BCE

Oldest evidence of harnessing of dogs in the Arctic, suggesting they were used for transport over distances of at least 1,500km (930 miles)

7,000 BCE

Date of oldest-discovered dog poo, found in a Chinese village

3,300–600 BCE

Bronze Age illustrations and cave paintings depict dogs

800 BCE

In Homer's *Odyssey*, Odysseus returns after 20 years away and is only recognized by his dog, Argos

1434

Van Eyck's *Arnolfini Portrait* features a small dog with mesmerizing eyes, representing marital fidelity. Weird.

1873

The Kennel Club established in UK, setting breed standards

2.02 Is your dog basically a cute wolf?

Dogs share 99.96% of their genes with wolves, and some breeds look very similar to wolves. So, is your own dog a ferocious blood-thirsty wolf wrapped up in a cute little pooch-like package? If you set her free, would she return to the mountains and spend her nights howling at the moon and running free with the pack?

Almost certainly not. Living with humans has profoundly changed dogs' needs and lifestyles and this has shaped their physical and intellectual abilities, and how they behave, function, reproduce and socialize. The first dogs to be domesticated would have had an unusual combination of fearlessness and friendliness towards humans. These traits have since intensified because humans only kept and bred friendly, useful dogs. After all, no one wants a vicious baby-eating predator hanging around the cave, do they? So, what, exactly, has changed?

Behaviour

Dogs bark but wolves rarely do. Wolves howl but dogs rarely do. Dogs are playful, even into adulthood, and form stronger bonds with humans than with other dogs. Domestic dogs have been chosen for tameability, adaptability and for their extraordinary ability to read human communicative gestures. Unlike wolves, they are also dependent on humans, rather than each other, for food. Research has shown that wolves cooperate to solve problems in order to get food, but dogs rarely do. Wolves are fearful and aggressive towards humans, and although **wolf pups can be socialized, they can't really be domesticated. Even if raised with humans from**

birth, they never become as close to people, read our body language, or look at us as often as dogs do.

Pack life

Wolves are wild animals best suited to living and hunting large mammals in complex social packs, protecting themselves and their offspring against predators. Grey wolf packs usually contain five to 10 individuals: an adult alpha male and female breeding pair, plus their offspring and some unrelated wolves. They work together to hunt prey and raise their young, have a clear social structure and rules of conduct, and are extremely loyal to each other. Alpha wolves are usually the only ones in the pack to mate and will ensure that pups get to eat before others finish everything. In contrast, dogs aren't even seen as pack animals any more. Groups of feral dogs (domestic dogs that live in the wild) are scavengers rather than hunters: they don't cooperate to feed; often quarrel; mate with anyone, including their own relatives (which is generally bad for genetic diversity); raise their young alone; and don't stay in fixed family packs.

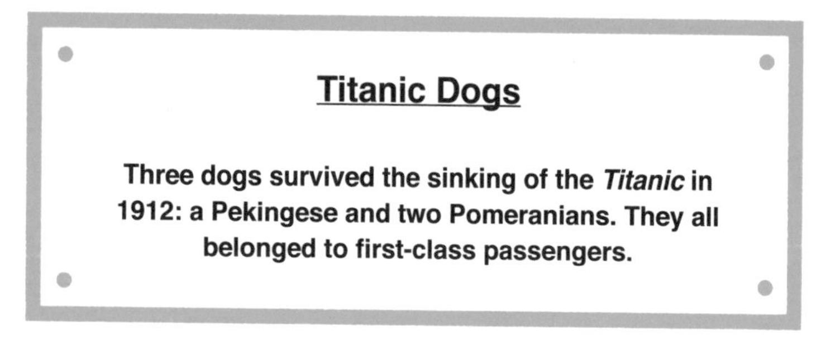

Titanic Dogs

Three dogs survived the sinking of the *Titanic* in 1912: a Pekingese and two Pomeranians. They all belonged to first-class passengers.

Diet

Wolves eat mainly herbivorous ungulates (hoofed animals) hunted by the whole pack, but will eat smaller prey and even insects when food is scarce. They consume a negligible amount of vegetable matter and are true carnivores. Dogs, on the hand, are omnivorous – they can eat a diet that includes vegetable matter, such as grains, a quirk thought to have developed as a result of eating human leftovers. That said, dogs do require some nutrients that would, in the wild, only come from meat.

Self-reliance

As they can't depend on humans to help them, **wolves are highly self-reliant and will tackle a problem such as a closed door by trying to open it themselves. When dogs meet a tricky problem, they often look for a human to solve it for them.**

Reproductive cycle

Female wolves always mate with the same male and give birth once a year in spring to ensure the best chance of survival for pups and the whole pack. Spring is when food becomes abundant, and offers the maximum amount of time between the arrival of new pups and the onset of winter, when food grows scarce. By contrast, female dogs mate with multiple partners and have two breeding cycles a year that can be in any season. Because dogs depend on humans for food and shelter, there's no optimal time of year to ensure survival of their young.

2.03 How were dogs domesticated?

We're not exactly sure when, where or why dogs were first domesticated (in fact, we're not even sure whether humans domesticated dogs, or if dogs domesticated themselves). They were the only animals to be domesticated before settled agriculture began around 10,000 years ago, and it's pretty clear that at some stage both dogs and humans benefited from the deal. Dogs got a steady supply of food, shelter, protected reproduction and companionship, and **humans got a hunting, herding, hauling, bed-warming, pest-chasing security alarm, and a source of food and even wool that also gave them companionship.**

Canine domestication can't have happened before 40,000 years ago (the earliest estimated date of dogs' divergence from wolves) but there's little archaeological evidence until the famous Bonn-Oberkassel dog (discovered in 1914), which was buried with a human couple over 14,000 years ago. That dog was 28 weeks old and had been gravely ill with canine distemper from 19 weeks – it must have been cared for by humans to have survived as long as it did.

A 2021 literature review study published in the journal *Proceedings of the National Academy of Sciences* suggests that the dog was domesticated in Siberia 23,000 years ago, but there's no archaeological evidence to support this. Some studies indicate that domestication could have started over 25,000 years ago, while a 10-fold increase in the dog population from around 15,000 years ago may mark when dogs fully benefited from domestication. Some convincing research from the University of Oxford in 2016 even suggests that dogs may have been domesticated twice – once in the East and once in the West.

Along with the domestication of farm animals, the human-canine bond is thought to have been a key factor in our species' development, helping to achieve our transformation from hunter-gatherers to settled farmers.

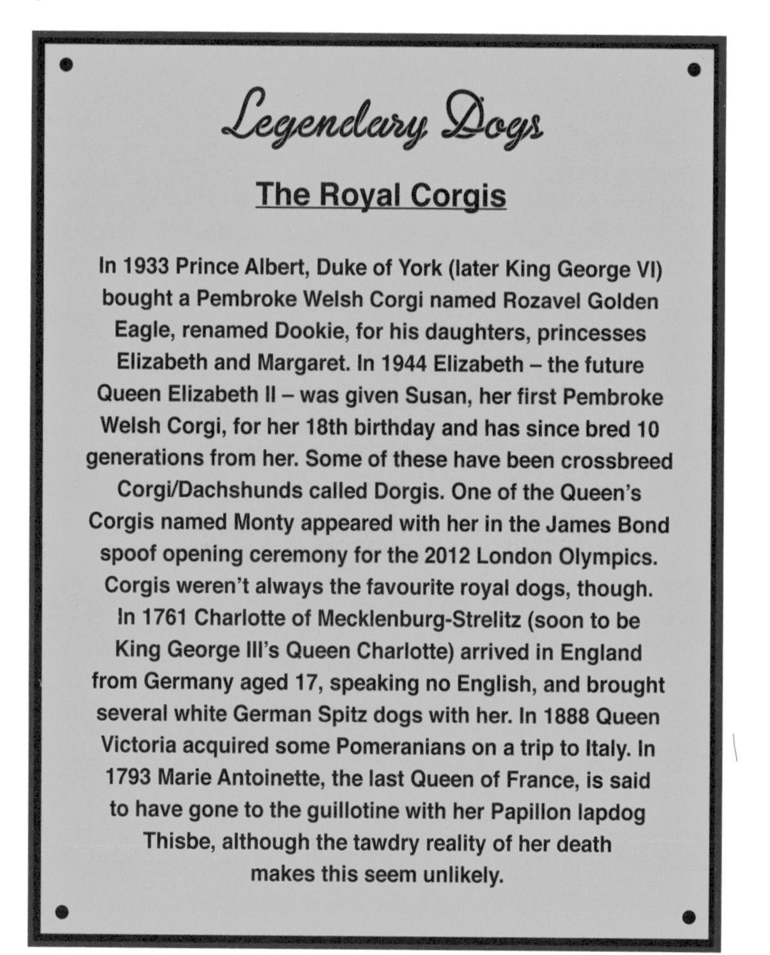

Legendary Dogs

The Royal Corgis

In 1933 Prince Albert, Duke of York (later King George VI) bought a Pembroke Welsh Corgi named Rozavel Golden Eagle, renamed Dookie, for his daughters, princesses Elizabeth and Margaret. In 1944 Elizabeth – the future Queen Elizabeth II – was given Susan, her first Pembroke Welsh Corgi, for her 18th birthday and has since bred 10 generations from her. Some of these have been crossbreed Corgi/Dachshunds called Dorgis. One of the Queen's Corgis named Monty appeared with her in the James Bond spoof opening ceremony for the 2012 London Olympics. Corgis weren't always the favourite royal dogs, though. In 1761 Charlotte of Mecklenburg-Strelitz (soon to be King George III's Queen Charlotte) arrived in England from Germany aged 17, speaking no English, and brought several white German Spitz dogs with her. In 1888 Queen Victoria acquired some Pomeranians on a trip to Italy. In 1793 Marie Antoinette, the last Queen of France, is said to have gone to the guillotine with her Papillon lapdog Thisbe, although the tawdry reality of her death makes this seem unlikely.

2.04 Why do humans love dogs?

Dogs are immensely useful to us, but simply being useful doesn't equate to love. For example, my ergonomically designed Bosch PSB 1800-volt LI-2 cordless combi drill in British racing green is useful to me, but do I love it? Actually, when I think about it, I probably do. A better example would be my set of drop-forged adjustable spanners in four different sizes. Do I love them? Okay, maybe not a good example, either. But my point is this: just being useful isn't enough. You've got to *feel* something, too.

The reason for our love of dogs is biochemistry. **When we interact with dogs our bodies release the hormones oxytocin, beta-endorphin, prolactin and dopamine, and the neurotransmitter beta-phenylethylamine – all associated with affection, happiness and bonding.** We also experience a decrease in cortisol, a hormone related to stress. Put simply, when we pet dogs, we feel a pleasant biochemical high and that high, together with the cycle of bonding, nurturing, attachment and companionship it creates, is as good a definition of love as any. And the fact that I feel all those things from a beautifully engineered, top-of-the-range power tool doesn't make that love any less important, whatever my wife says.

There's also solid evidence that **when we stare at dogs they activate the same hormone release we get from staring into the eyes of babies**. In a sense, dogs have hijacked the biochemical system that binds humans together. Compare this to other pets: I've never had a hormone rush from staring at a goldfish, although I might have felt one from Gerry the Gerbil when I hit a confusing patch at the age of 11.

The simple act of nurturing (caring for another) has a positive impact on humans. Research shows that when people aren't able, or aren't allowed, to care for others, they are more likely to experience decreased health, wellbeing and depression. Caring for a dog makes us feel good.

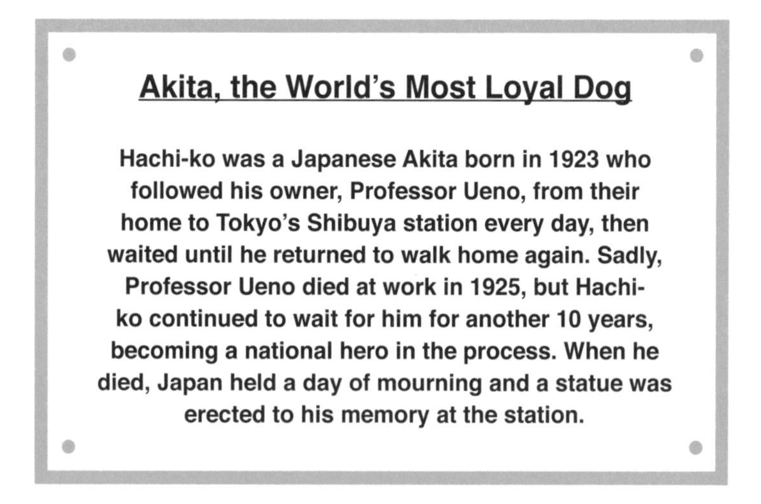

Akita, the World's Most Loyal Dog

Hachi-ko was a Japanese Akita born in 1923 who followed his owner, Professor Ueno, from their home to Tokyo's Shibuya station every day, then waited until he returned to walk home again. Sadly, Professor Ueno died at work in 1925, but Hachi-ko continued to wait for him for another 10 years, becoming a national hero in the process. When he died, Japan held a day of mourning and a statue was erected to his memory at the station.

2.05 Why do dogs love humans?

The simple answer is that they don't have a choice. We humans only bred the dogs that happened to love us. And we've been remarkably effective: research shows that most dogs love people even more than they love other dogs.

But why are dogs so much more loving than other animals? One fascinating reason may be found in their genes. Geneticist Bridgett vonHoldt at Princeton University discovered evidence of **a genetic quirk caused by domestication that makes dogs hypersocial** and, specifically, much friendlier than wolves. Stick with me here, because this is a little complicated. The quirk is a section of disrupted DNA (on a gene for a protein called GTF21), and this section can be disrupted in different ways, including in the degree of variation in the disruption. The more disrupted the section, the friendlier the dog is; the less disrupted, the more aloof and wolfy it is. There's a very close association with the human congenital disorder known as Williams-Beuren syndrome, which (among other things) makes people unusually trusting and friendly. Changes in the same gene in mice cause them to be hypersocial, too. In contrast, undisrupted wolf DNA seems to make them more aloof and wary of humans. It's possible that as we selected the friendliest dogs, we selected those with a canine equivalent of Williams-Beuren syndrome, and the hypersociability may have become hereditary.

Allowing wild animals into the home would have been a big risk for our ancestors. It meant sharing food and compromising the safety of children. Thus they would have only bred animals that were calm, protective of their adopted family, and useful to have around. Friendly dogs would have gained most from positive interactions with

humans, which brings us back to biochemistry (see p20). Just as we get a pleasurable neurotransmitter release from interacting with dogs, **dogs get the same from interacting with us: the hormones oxytocin, beta-endorphin, prolactin and dopamine, and the neurotransmitter beta-phenylethylamine – all associated with affection**, happiness and bonding. They don't experience the same decrease in the stress hormone cortisol as we do, but they still get a similar pleasant biochemical high.

Chapter 03: Doganatomy

3.01 Dog sex

Oh great. As if it wasn't bad enough to be teaching canine reproduction on a Monday morning, fate has handed me 9G to teach it to – the class whose immaturity defies Darwin's laws of evolution.

Right, sit down and let's get this over with. If you want cute little puppies in the world, dogs are going to have to have sex and give birth. Most female dogs become sexually mature at between six and 16 months of age, meaning that they have grown all the necessary equipment to reproduce, and their bodies can produce hormones to start ovulation. Most males become mature at around 10 months, something I suspect the boys of 9G may never achieve. This is very different to the dog's closest ancestor, the wolf, which doesn't mature until around two years of age. Wolves also tend to be monogamous, whereas dogs will welcome multiple partners – and no, Gary, there are no comparisons whatsoever to be drawn with Mr Norris from the chemistry department.

Female dogs come into season twice a year (when they are able to get pregnant and have puppies), whereas wolves only come into season once a year. This is likely because humans selectively chose dogs that were able to breed more often.

When it comes to mating, male dogs show the female a lot of interest, sniffing her and dancing around while lowering their front half and wagging their tail. And no, Gary, I don't think you did spot Mr Norris doing that to Mrs Jones at the Christmas disco. The male will often nip the female around the face, neck and ears, and jump up at her from the side. The female is in charge, and if she doesn't like the male, she'll bite and growl at

him, or simply roll over. If she finds the male attractive, she'll look submissive, make whimpering noises and pull her tail to the side.

Right, buckle up, here we go ... The dog's penis has two fascinating quirks: firstly (and very rarely for the animal kingdom), it contains a thin bone called the baculum, which keeps it straight when he mounts the female and inserts it into her vagina. Second, after insertion, an extra lump near the base of the penis called the bulbus glandis expands and effectively locks the two animals together. The male ejaculates sperm that, if mating is successful, will reach the female's eggs and fertilize them. **The male then dismounts but, because of the expanded bulbus glandis, the animals stay joined together for a surprisingly long time – between five and 80 minutes.** And the weirdest thing is that no one is entirely sure why. Ah Gary, it looks as though you've drawn some very detailed diagrams of the whole process. Perhaps you'd like to show them to the class and explain why the one with the enormous bulbus glandis looks remarkably like Mr Norris?

Pregnancy lasts 60–68 days (compared to humans' 280 days), and the average litter is between six and eight puppies, although anything from one to 14 is normal. Wonderful though all this is, it's a good idea to neuter dogs to prevent unwanted puppies that may end up being euthanized because they can't be homed. And when it comes to 9G, the temptation is strong.

Well, thank God that's over. I'm off to the science department to see if Mr Norris has any ethanol left over from his distillation project.

3.02 Do dogs sweat?

Not really. They have a few eccrine sweat glands (glands that open directly on to the skin surface) on their paws, but not enough to help them regulate their temperature on any great scale. Regulating the body's systems to keep everything in balance is called homeostasis, and it combines breathing, circulation, energy, hormone balance and body temperature regulation.

Dogs' normal body temperature is around 38.5°C (101.3°F) – a good 1.5°C (2.7°F) higher than ours at 37°C (98.6°F). When humans get too hot we thermoregulate (control our temperature) by sweating, breathing, and via simple radiation from our skin. Dogs, however, are mostly covered in a thick insulating layer of fur so sweating and radiation would be downright dangerous. **If dogs did sweat, they would quickly turn into heavy, humid, fetid wet mops crawling in parasites and with a revolting smell**, having created the perfect environment for bacteria to thrive.

By far the most important thermoregulating tool dogs use is panting, and it cools them in much the same way that sweating does with humans, except in dogs it happens inside the body. Dogs have a surprisingly large surface area in their nasal cavities, mouths and tongues, which are kept damp by saliva and are rich in capillaries carrying warm blood close to the surface. As dogs pant, air passes over these damp surfaces and cools the blood underneath by evaporative heat exchange. While the tongue is rich in capillaries, the nasal cavity is the most effective, so when a dog starts panting through the mouth its thermoregulation has moved into top gear. Vasodilation can also help dogs stay cool – the blood vessels in their face and ears expand, helping to increase heat radiation, and in summer they shed their soft, insulating undercoat hairs to make this more efficient.

3.03 Why do dogs poo facing north–south?

In one of the strangest yet fascinating studies ever published, Czech and German researchers found that **dogs prefer to align their bodies north–south to poo, guided by the Earth's magnetic field**. Female dogs also align like this to pee, but not males (raising their leg seems to interfere with the alignment). The 2013 study published in *Frontiers in Zoology* followed 70 dogs over two years, making 5,582 observations, and not only did it prove that this magnetosensitivity existed, it showed dogs were intensely sensitive to it. The Earth's magnetic field fluctuates, moves and can even flip (yup: Earth's north/south poles have reversed polarity in the past), and whenever it's unstable, this directional behaviour in dogs pauses.

I know, 'mindblowing' doesn't even begin to describe it. It turns out that this behavioural quirk is well-known in grazing and resting cows and deer. Red foxes hunt using magnetosensitivity and are more successful at catching mice when they pounce on them in a north-east direction (I promise you I'm not making this up).

If you thought it couldn't get any more fascinating, a 2016 study published in *Nature* identified cryptochromes in the photoreceptors in dogs' eyes. These light-sensitive molecules are one of the tools that birds can use in daylight to navigate by light-dependent magnetic orientation (meaning they only react to the Earth's magnetic field when they are also excited by light). This raises the possibility that dogs can see the Earth's magnetic field, although more research is needed to be sure. It might sound far-fetched, but bear in mind that other animals have sensitivities way beyond ours: some sharks use a hypersensitivity to electrical charge to hunt their prey; many insects and fish can see ultraviolet light; and many hunting snakes have infra-red vision.

3.04 How many hairs are there on your dog?

Every dog owner wants to know how many hairs their dog has, don't they? It's a tricky one to answer: selective breeding has produced a huge variety of sizes and shapes of dog, some with a single coat type, others with double coats, some with spectacular drapes of dreadlocks (the Komondor), and others with no hair at all (the Mexican Hairless). But let's give it a try.

First, we need to work out the body surface area of your dog using their weight. This can be tricky as the surface area-to-weight ratio varies dramatically between small thin breeds and large rotund breeds, but luckily the *MSD Veterinary Manual* has a handy online conversion table. Generally speaking a little 5kg (11lb) dog has a surface area of 0.295m^2 (3.175 sq ft), a 10kg (22lb) dog has one of 0.469m^2 (5.048 sq ft), a 20kg (44lb) dog one of 0.744m^2 (8.008 sq ft), a 30kg (66lb) dog 0.975m^2 (10.495 sq ft) and a 40kg (88lb) dog 1.181m^2 (12.712 sq ft).

Once you've worked out the surface area, you'll need to multiply that by the average amount of hairs on a dog per cm^2 (0.155 sq in). According to *Miller's Anatomy of the Dog* **there are an average 2,325 hairs per cm^2 on a dog**, so multiply that by the surface area of your dog and VERY roughly, that works out at:

Breed	Av. Weight	Surface area	Hairs
Miniature Dachshund	5kg (11lb)	0.295m^2 (3.175 sq ft)	685,875
French Bulldog	10kg (22lb)	0.469m^2 (5.048 sq ft)	1,090,425
Cocker Spaniel	14kg (31lb)	0.587m^2 (6.318 sq ft)	1,364,775
Border Collie	17kg (37lb)	0.668m^2 (7.190 sq ft)	1,553,100
Labrador/ Golden Retriever	30kg (66lb)	0.975m^2 (10.494 sq ft)	2,266,875
Rottweiler	49kg (108lb)	1.352m^2 (14.552 sq ft)	3,143,400
Great Dane	60kg (132lb)	1.560m^2 (16.791 sq ft)	3,627,000

Dreadlocked Dogs

Hungarian Pulis have thick, greasy coats that develop naturally into dreadlocks. I highly recommend an internet search for 'Hungarian Puli running'. The fur of Komondors, a similar breed and also Hungarian, looks similar to the sheep they are bred to protect. The sheep aren't afraid of them and the dogs grow up feeling protective of their herd 'family'. The disguise is ruined every spring when both sheep and dogs are shorn, which must be quite a shock for everyone.

3.05 Why is your dog so damn CUTE?

Canine cuteness and evolutionary science are connected to a surprising degree. Dogs have come a long way from all those scary-looking wolf relatives with their erect ears, large body size, sly-looking eyes and long snouts. With the exception of certain breeds, dogs have evolved floppier ears; larger, rounder eye structure; a shorter snout; a compact body and an extraordinary doleful expression designed *specifically* to make us weak at the knees.

This expression comes from **an ingenious muscle that seems to have evolved to snare humans. It's called the levator anguli oculi medialis (or LAOM), and its function is to make dogs look cute, sad and doleful**. It sits above the eyes, close to the centre of the forehead, and when tensed gives a dog's face that crinkled-forehead, huge-eyed look of sadness and vulnerability that taps directly into our well-researched desire to nurture. The LAOM muscle may have started life as a mere genetic quirk but it has become a powerful manipulative tool – research shows that rescue shelter dogs that deploy it when meeting strangers are more likely to be adopted. Wolves, by contrast, don't have this muscle.

In behavioural terms, humans chose gentle, tameable animals that are happy and confident around humans. More interestingly, **we've created a species that's rare in the animal kingdom because it thrives on human company and loves to play not just when young, but deep into adulthood**.

In physical terms dogs are very cute, with large eyes and shorter snouts than wolves, so you could simply conclude that humans have selectively bred dogs for sheer cuteness. But canine looks probably weren't particularly important to early humans who struggled to

survive, so how did this puppy-like big-eyed animal evolve? Well, there's a fascinating twist to domestication called neoteny. Basically, when you select dogs purely for friendliness rather than how they look, they also end up retaining juvenile attributes, even when mature.

In the 1950s, a Russian geneticist called Dmitry K Belyaev decided to re-create the domestication process to see how evolutionary changes came about by building a tame fox population using silver-black foxes, and this shed light on the idea of neoteny. The controversial project (there seems to be a lack of data surrounding the study) is still ongoing, although Belyaev died in 1985. He started with 100 vixens and 30 males – the friendliest he could find – and when cubs were born they were hand-fed by researchers but had minimal contact with humans. The friendliest 10% of each new generation were kept and the rest were sent back to the fur farms their parents had come from. Brutal stuff, but fascinating. **Within four generations the fox cubs became dog-like, wagging their tails, seeking out contact with humans and responding to gestures and glances**. They whined and whimpered, and licked the researchers like dog puppies, and the adults were playful and friendly. They also became sexually mature at an earlier age and could breed out of season, producing larger litters.

The strangest result of the study was that strong physical changes also emerged (even though the foxes have only been selected for tameability – not shape and size). The domesticated foxes seemed to have floppier ears, shorter legs, curlier tails, upper jaws and snouts, as well as a widened skull. All the traits that humans see as 'cute'. I should emphasize that the results of Belyaev's project have not been fully published or explained, and evidence around domestication syndrome is not yet clear, but it does seem as though domestication itself changes the physical appearance of some canines.

3.06 The science of paws and claws

Dogs are digitigrade, which means they walk on the tips of their toes, unlike humans who are plantigrade (we walk with toes and metatarsal bones flat to the ground), or cows and horses, which are unguligrades (they walk on the tips of the toes, usually with hooves covering them).

The pads on dogs' paws are made of keratinized epidermis (skin made from the tough protein keratin, like human fingernails and hair). Each foot has four digital pads (dogs' version of toes) crowning a heart-shaped metacarpal pad (like the heel of our palms), as well as a little-used dewclaw on the inside of their front legs and, more rarely, on their back legs, too (this is like a tiny human thumb). They also have a carpal pad on their forelegs that has no equivalent in human anatomy, but is occasionally used to help dogs stop fast when descending a steep slope.

Remarkable Lundehunds

Norwegian Lundehunds have a range of special party tricks: first, they are polydactyl, with double dewclaws, giving them six toes on each foot. They can also turn their heads 180°, turn their forelegs at 90° to their bodies, and fold their ears shut – both forwards and backwards! They were originally bred to hunt puffins.

Dewclaws are particularly odd: all dogs have them, but in many breeds they're so small they can't be seen. Front dewclaws often have a little bone and muscle in them, but rear dewclaws rarely have either. Some breeders even get them surgically removed, saying their lack of structure means they can get ripped off, causing great pain. Certain breeds have dewclaws on both front and back legs, and Great Pyrenees often have double dewclaws on their hind legs. They are occasionally used by dogs to get a better grip on bones or (in my dog's case) tennis balls.

The pads work as shock absorbers and they also have sweat glands that help dogs balance their body temperature, although aren't very effective at this (dogs sometimes perspire through their feet from stress or nervousness, too). **The distinctive smell of dogs' feet is due to the fungi and bacteria that thrive in damp paws packed with lots of microscopic cavities**. Dogs' claws can't retract in the way that cats' can and, unlike human toenails, are directly connected to the bone, containing nerves and blood vessels. Just one of the many reasons that dogs hate having them clipped.

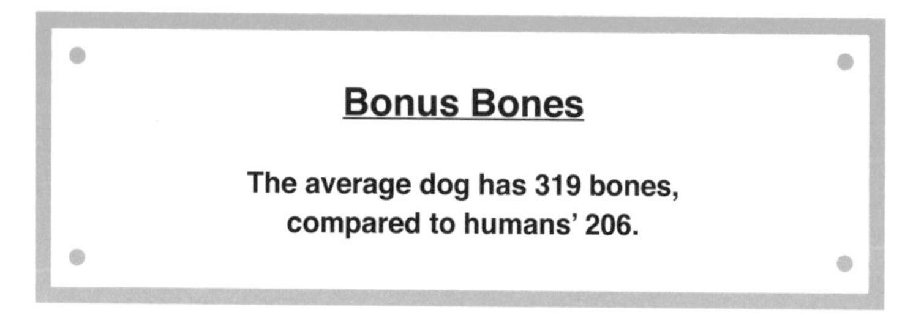

Bonus Bones

The average dog has 319 bones, compared to humans' 206.

3.07 Why do dogs make such a mess when drinking?

Dogs' drinking may look slapdash and messy, but it's actually a fascinating, highly technical, and precisely timed set of movements called acceleration-driven open pumping. The process looks similar to a cat's lapping, but is very different. Both cats and dogs' tongues move too quickly to properly observe with the naked eye, but a study published in the journal *Proceedings of the National Academy of Sciences* used high-speed photography to work out what's going on.

Dogs' mouths are designed to allow them to open their jaws wide so they can bite large mammals, and as a consequence they have incomplete cheeks. This means that unlike humans, horses and pigs, dogs can't create suction in their mouths to suck up water. Instead, they curl their long tongues backwards, creating a ladle-shape that thumps down into the water creating an initial splash. The tongue is then quickly yanked back into the mouth, and the splashed water sticks to the top side of the tongue, creating a rising column of water. The dog then snaps its mouth shut into this column of water at the point when the greatest volume has risen, and drinks it. The snapping-shut action causes a little bit of mess, but the whole process means that dogs are able to drink more per lap than they could with a straight tongue.

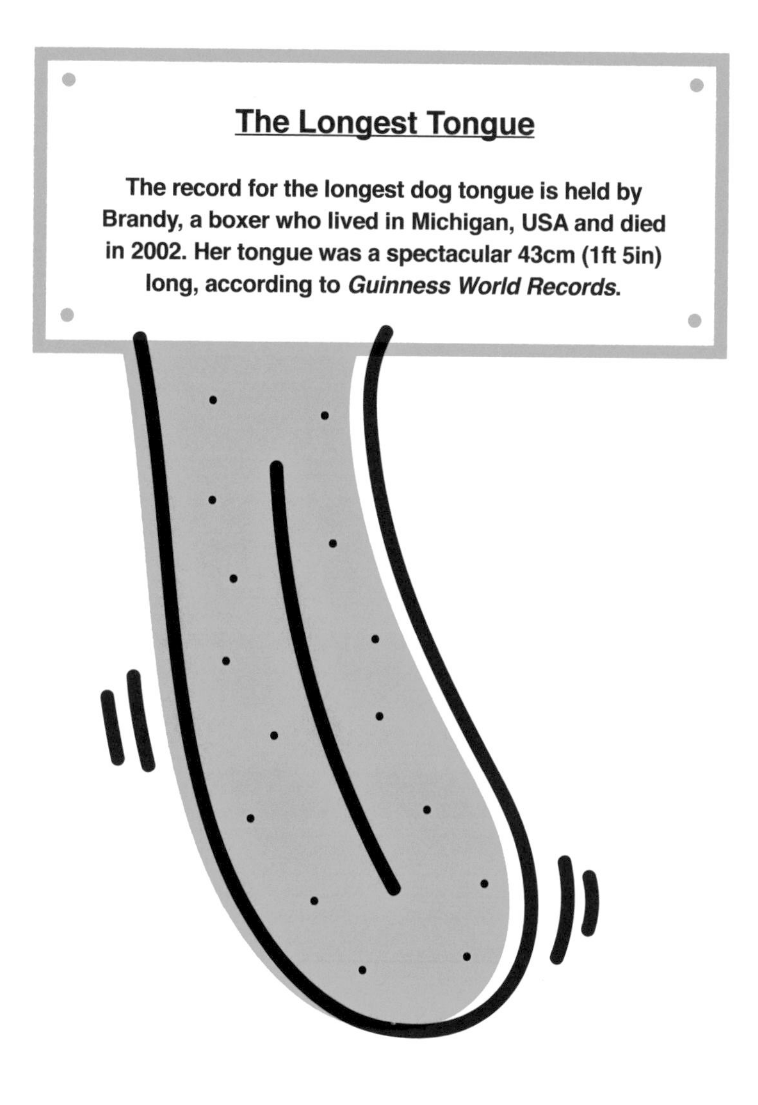
The Longest Tongue
The record for the longest dog tongue is held by Brandy, a boxer who lived in Michigan, USA and died in 2002. Her tongue was a spectacular 43cm (1ft 5in) long, according to Guinness World Records.

3.08 Is one dog year really equivalent to seven human years?

The median longevity of dogs (the age at which half are alive and half have died) is 10–13 years, depending on breed. Seeing as worldwide human life expectancy is 70–72 years, it seems obvious to simplify that into the popular formula of one dog year = seven human years. But the reality is much more fascinating.

Comparing the development of two wildly different animals is difficult but we can match certain life events: weaning (no longer relying on your mother's breast milk), sexual maturity, physical abilities and failing health. In old age, dogs suffer some of the same problems as humans: arthritis, dementia and joint conditions. Researchers have also matched life stages using a comparison of methylomes – chemical changes to genes that fluctuate throughout our lives.

It turns out that dogs mature extremely fast to begin with, but their development slows down so much by the age of 11 that in real time it's even slower than humans'. In the early years dogs age spectacularly fast, hitting sexual maturity at around six months compared to humans' 15 years, and full structural development by nine months. At the age of one, dogs are the human equivalent of 30 years old and by the age of three they equate to 50.

Of course there are exceptions to this. Mixed breed dogs live around 1.2 years longer than pure breeds, and smaller breeds live longer than larger breeds: mastiffs often only live to seven or eight years but Miniature Pinschers average 14.9 years.

3.09 How old is your dog in human years?

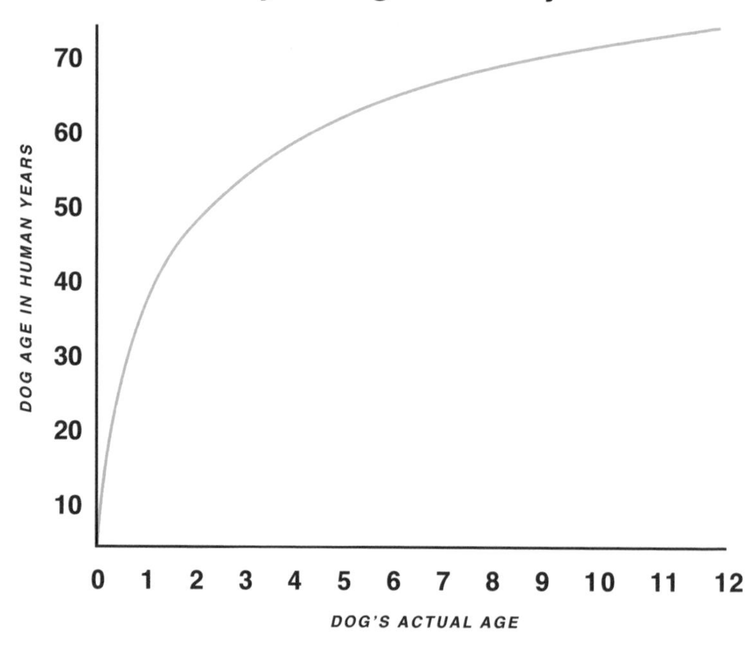

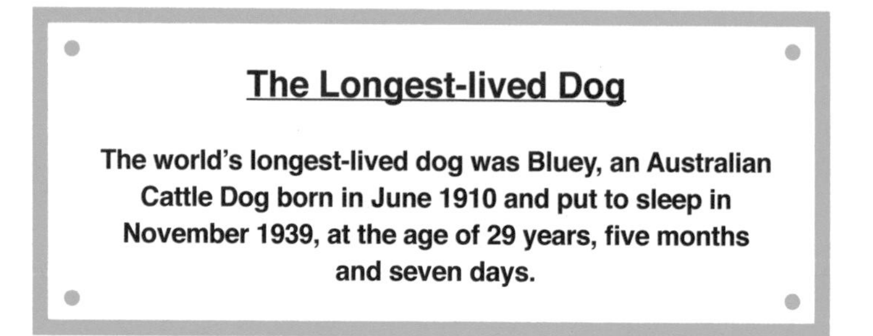

The Longest-lived Dog

The world's longest-lived dog was Bluey, an Australian Cattle Dog born in June 1910 and put to sleep in November 1939, at the age of 29 years, five months and seven days.

Chapter 04: Revolting Doganatomy

4.01 Why do dogs fart (but cats don't)?

Dogs are omnivores and eat pretty much anything they can fit in their greedy mouths. They do require some micronutrients that in the wild they would only be able to get from meat, but they have a digestive system designed to cope with everything, including fibre.

On the other hand, cats are obligate carnivores, which means their digestive system is designed for a 100% meat diet: rich in protein and fat but very low in carbohydrates or fibre. Cats have a shorter digestive tract than most animals and their digestion is focused on breaking down protein and fats into smaller molecules. (Incidentally, cats don't create glucose sugars from carbohydrates in the way that humans do. Instead they produce it by gluconeogenesis in their livers. This breaks down proteins first into amino acids, then further into glucose. Amazing, isn't it?) So, **the main digestive difference between cats and dogs is that dogs are built to break down fibre and cats aren't**.

What does this have to do with farts? Well, dog farts are mainly produced from fibrous foods such as grains and veg, as these are broken down by bacteria in the colon via fermentation rather than by enzymes in the small intestine. But the by-product of bacterial fermentation is gas – actually, lots of different gases, and some of them are very stinky indeed. Because cats don't eat fruit and veg, they have very little fibre in their diet for any bacteria to ferment, and hence very little fart. A fair proportion of dogs' farts are also made of air gulped while eating – dogs tend to wolf food down, swallowing lots of air, whereas cats tend to eat daintily and slowly.

Dogs eat anything, fart at will and often look pretty pleased with themselves after doing so. Much like me.

4.02 The science of dog poo

Dog poo (or 'canine faeces' if you're applying for research funding) crops up in some extraordinary places. One sample was recovered from a 7,000-year-old Chinese farming village and another was found in a 17th-century British chamber pot. Clearly some lazy sod from 400 years ago had taught their dog to poo in a pot to avoid taking it out for a walk. **UK dogs produce an estimated 365,000 tonnes (400,000 tons) of poo a year (by comparison the Empire State Building weighs 331,000 tonnes/365,000 tons)**. In an attempt to solve their city's poo problem, residents of New Taipei City in Taiwan were offered a lottery ticket for every bag of poo they handed in. In total 14,500 bags were collected from 4,000 people, and a woman in her fifties won a gold ingot worth $2,200 (£1,400). The scheme is thought to have halved the amount of excrement in the city.

So, what's in dog poo? Well, that depends on your dog's diet, age and state of health, but generally it contains billions of bacteria (both live and dead), undigested food (especially fibrous food) that hasn't been broken down, old cells shed from the digestive system, as well as any of the marvellous juices, enzymes, biles, acids and other secretions that the body produces to break food down that haven't been re-absorbed through the intestinal wall. There are also gases, short-chain fatty acids and a few other bits and bobs, but none of these are responsible for the delicious aromas of a good poo. No, those come from the hydrogen sulphide, indole and skatole produced when undigested protein makes its way through to the large intestine. Only tiny amounts of these chemicals are produced, but they do pack a powerful honk. Interestingly, when dogs poo they also add smelly pheromone

secretions from their anal sacs (glands on either side of their bum) that leave behind information about their age, sex and identity.

Let's look at the numbers: **9 million tonnes (10 million tons) of dog poo is produced every year in the US alone, with surveys showing that only 60% of it is picked up and disposed of**, which is crap, really. But even when it's put in a compostable bag, dog poo can still be problematic. If that bag is thrown into a general rubbish bin, it will end up in landfill, won't compost properly and will ferment to produce methane – a particularly problematic greenhouse gas. The best thing would be to add it to compost (although it is pretty smelly and needs to be carefully controlled). Poo is potentially useful and beneficial to the environment, but only when treated properly, so this is an area where a solution is desperately needed.

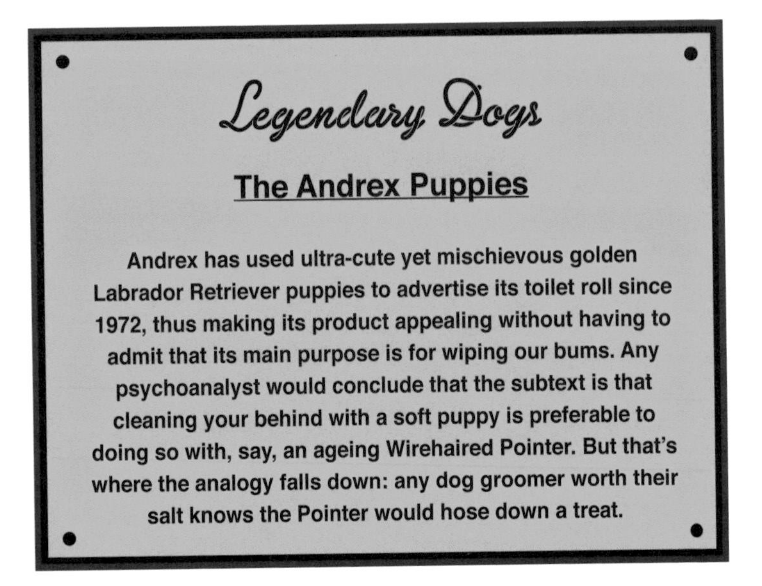

Legendary Dogs

The Andrex Puppies

Andrex has used ultra-cute yet mischievous golden Labrador Retriever puppies to advertise its toilet roll since 1972, thus making its product appealing without having to admit that its main purpose is for wiping our bums. Any psychoanalyst would conclude that the subtext is that cleaning your behind with a soft puppy is preferable to doing so with, say, an ageing Wirehaired Pointer. But that's where the analogy falls down: any dog groomer worth their salt knows the Pointer would hose down a treat.

4.03 The science of dog pee

Dog pee is remarkably similar to human pee, containing **mainly water (95%), with a surprisingly large amount (5%) of organic and inorganic waste material, metals and ions dissolved in it**. I once built urine from scratch for a BBC TV programme, using all the individual ingredients. It was quite explosive. Let's just say: it's definitely not a good idea to drop potassium into water without a lot of safety precautions.

Urine is basically a tool for flushing substances out of the body, especially nitrogen-rich by-products of cellular metabolism (the process our bodies' cells use to produce and consume energy). These include the organic compounds urea, creatine and uric acid, as well as carbohydrates, enzymes, fatty acids, hormones, inorganic ammonia, chloride ions and the metals sodium, potassium, magnesium and calcium. Lots of these substances are used in the home: urea is often sold as a de-icing powder and is an ingredient in hair-removal cream, animal feed and moisturizing cream; while creatine is used as a food supplement for people with muscle problems and as a performance-enhancer for athletes.

Dogs are fascinated by each other's pee because it contains smelly substances such as pheromones that contain lots of information. These communicate if a dog is male or female, its reproductive stage, its age, emotional state and even if it has a disease such as diabetes. Males especially like to mark territory by peeing wherever they fancy, although contrary to common perception, this marking seems to be more about saying hi than warning other animals away.

4.04 Why do male dogs raise their legs to pee?

The day that your male puppy first raises his leg to pee up a lamppost is a day of mixed feelings. On the one hand, you feel like a proud parent of a furry little boy taking small steps towards maturity, and on the other, there's the dawning realization that from now on the little bugger's going to cock his leg up everything and everyone, and the indiscriminate humping stage probably isn't far behind.

But why do males have to lift their legs so publicly, whereas females squat so daintily? Well it's partly because males can. The bony baculum in their penis holds it straight and makes peeing more directional, allowing them to lift their leg, point Percy at the porcelain, and hit a spot, without causing too many personal hygiene problems. If female dogs do the same (and it's not unheard of), they're more likely to spray some pee on themselves, which can end up causing infections or damaging fur.

So, given the tool to do it, dogs take the opportunity to use their pee to spread the word about themselves by scent marking (urine contains lots of information about dogs' sex, health and age, see p54). But there's a fascinating twist to this: a study published in the *Journal of Zoology* showed that **small dogs tilt their legs higher against objects than large dogs do, possibly using it as an opportunity to deceive by exaggerating their size and competitive ability.**

4.05 Hangers-on: what are fleas, ticks and mites?

Fleas

The most common flea found on dogs is, ironically, the cat flea, *Ctenocephalides felis*. These are 2–5mm (0.08–0.2in) long with six legs, and a laterally flattened body (they look like they've been squished in the doors of a lift). They're capable of jumping around 22cm (9in) – that's equivalent to a human jumping 90% of the height of the Empire State Building.

Fleas feed exclusively on the blood of their host animals. Their life span is anything from 16 days to 21 months and they can live up to a year without food in the right conditions. In fact they spend most of their life cycle off their host animal and only drink blood once they're adults. Once they've had a blood meal, they mature, reproduce and die within a few weeks. **Female fleas can lay up to 50 eggs per day, which soon fall off the dog. When they hatch into larvae they feed on flea excrement (yes, fleas live on their parents' poo)** while burrowed into rugs and bedding.

Fleas may be disgusting, but you have to admit they're pretty amazing. The downside is that they cause itching, blood loss, inflammation and allergic dermatitis. Get rid of them if you can – and preferably don't let them get a hold in the first place!

Ticks

Ticks are eight-legged arachnids and potentially very dangerous because they cause a variety of diseases (including Lyme disease, which is particularly nasty), allergies, anaemia, severe blood loss and tick paralysis. Adults are 3–5mm (0.12–0.2in) long depending on how engorged with blood they are, but the eggs, larvae and nymphs are

a fraction of that. They tend to be picked up in spring and summer when dogs brush against vegetation. They then crawl around the dog, usually making their way towards the head, ears or neck. Ticks need to feed on the host animal three times to complete their life cycle – and the females are easiest to identify as they're bigger when full with blood. Each female tick can lay up to 5,000–6,000 eggs before dying.

Dogs should be checked regularly for ticks (especially in spring and summer). These will feel like small mole-like bumps on their skin, and you may need to search through their hair to identify the revolting little things, usually with their heads and most of their legs buried, and only their engorged body and a few back legs visible. They should be killed with a topical insecticide, and then pulled out with tweezers or a little plastic tool made for the job.

Mange mites

The three main types of these horrible mites are the sarcoptic mange mite (which causes highly contagious canine scabies), the demodex mite (causing non-contagious demodicosis mange), and the ear mite, which infests dogs' external and internal ear canal.

None of them are very nice, and at less than 0.5mm (0.2in), most are too small to see, although a vet might just be able to spot the ear mite in an infected dog's ear wax. By far the worst are the sarcoptic mange mites, which can also cause infections in humans, and are hard to find as they burrow deep into the skin, producing toxins and allergens that inflame and irritate dogs, making them scratch, rub and bite. On the other hand, **demodex mites live quite normally in the hair follicles of most dogs and rarely cause any problems unless their population becomes too large.** They can cause dogs to have a moth-eaten appearance in patches of their fur, and lead to them needing an unpleasant series of insecticidal dips.

4.06 What are eye bogies?

The membranes inside the eyelids are called conjunctivae and they ooze a thin mucus called rheum (this is different to tears, which are produced by the lacrimal glands and are more watery and useful for washing away irritants). **Rheum is a slimy, water-based secretion that contains lots of substances, including antimicrobial enzymes to help prevent infections; immunoglobulins to identify viruses, bacteria and foreign objects; antibacterial inorganic salts and glycoproteins, all held together by mucins, which turn this whole marvellous soup into a viscous gel.**

This rheum is great stuff, helping to keep eyes healthy and soft, while repelling microscopic intruders. It's produced constantly and is usually washed away by watery tears whenever the eyelids blink. While dogs (and, indeed, humans) are asleep, tear-production is reduced and excess rheum will seep out of the eyes and dehydrate (the water in it evaporates), leaving behind the crust containing the other ingredients that were suspended or dissolved in the gel.

The same process happens in human eyes – and also in our noses when nasal mucus (also known as snot, and which contains many of the same components as rheum) dries out, leaving nice crispy bogies behind. So, calling this crusty rheum 'eye bogies' is pretty much spot on.

A few eye bogies are quite normal and can be removed with care (watch out – this is a sensitive area of your dog's face), but if your dog starts to produce a lot more than normal, or it's mixed with pus, you might have a mucopurulent discharge on your hands and possibly a condition such as allergic conjunctivitis. Off to the vet with you both.

4.07 Dog breath

Dogs' breath is highly variable, and almost always related to their oral health. That said, my least-favourite breath flavour has naff all to do with oral health and comes when my dog Blue licks me immediately after he's eaten cat sick.

Bad breath in dogs is pretty similar to that in humans: it's called halitosis, but the word refers to the symptom, not the cause. It usually results from plaque, tooth or gum disease and a build-up of bacteria at the back of the tongue. Plaque is a sticky deposit made from bacteria and fungi that grows on teeth, and can cause tooth decay as the bacteria break down sugars and create acid, which eats away at tooth bone. This plaque biofilm can also trap a build-up of anaerobic bacteria beneath it, causing inflammatory periodontal disease affecting gums, connective tissue and bone. Another common oral disease is gingivitis, which is an inflammation of the grooves between tooth and gum, and can also contribute to halitosis.

Bacteria love to make their homes in the microscopic crevices on the surface of teeth, so the solution to most dog breath is to brush their teeth with a specialist toothpaste. Other solutions are special (and expensive) dental diet foods alongside the dental processes of scaling (which removes plaque) and polishing (to smooth out microscopic crevices).

4.08 Why do dogs lick their genitals so much?

All dogs lick their genitals as part of normal grooming. It's not particularly pleasant for us to witness (especially when they try to lick our faces immediately afterwards), but it's all about hygiene. Without access to loo roll, shower gel and hot-and-cold running water, we'd probably end up doing the same. They lick after defecating and urinating to keep the area clean (the risk of catching a disease from licking the poo and wee is presumably lower than the risk of infection to the area they have licked).

But, let's talk mucous membranes. These are found on parts of the body that need to stay moist, usually bodily openings such as eyes, nostrils, mouth, anus, penis or vulva, and reproductive tract. **They ooze mucus, a useful gloop that helps to kill any bacteria, yeasts and viruses that try to make their way into the body**. But dogs sometimes need to clean up excess mucus, sweat, excretions and discharge from their various membranes. Sometimes these glands produce a little more gloop than is necessary and it needs to be licked away for hygiene purposes. It's perfectly normal, even if owners find it disgusting.

If the licking seems extreme, it could be due to a medical problem such as a urinary tract infection, allergy, skin infection or problem with their anal glands. If so, call that vet.

4.09 Is it bad to let a dog lick your face?

Why do dogs like licking faces? It's often a mark of affection – wolves do it to welcome a returning individual back to the pack, and dog puppies do it to increase the bonds between them. It also serves simple hygienic grooming purposes.

But licking has other uses, too: in the wild, female wolves eat food while out hunting and on their return puppies lick their face to encourage them to regurgitate it for them to eat. In packs of dogs, licking serves as a method of communication between the lower-ranking and more dominant members of the pack. The submissive dog will show that they know their place by crouching and licking the dominant dog, which stands tall and accepts the licking but doesn't respond. It might sound like a useless show of weakness, but it strengthens the pack's complex structure when everyone knows their place.

Then again, your dog may lick you because of positive reinforcement: because you responded positively, with smiles, laughter and possibly cuddles, when they licked your face in the past, they now repeat the behaviour to get more of the same. It's not too much of a stretch to see a lick as a kiss.

But what's in a dog's kiss? Well, a whole messy world of bacteria, viruses and yeasts for starters. Of course, our own mouths harbour a fair few microbes, but our **pets' mouths will include microbes from whatever else they've been licking: poo, dirt, their bottom, other dogs' bottoms, their genitalia, other dogs' genitalia** ... you get the picture. Dog saliva (like human saliva) also contains antimicrobial compounds that clean and heal wounds, but some will be unique to dogs, and our immune system may not be able

to deal with them. A study published in *PLOS ONE* showed that only 16.4% of the microbes it identified in dogs' mouths are shared between dogs and humans. However, **most microbiologists would say that you shouldn't be scared of a diverse microbiome** – most microbes aren't harmful and many are beneficial. But that's not to say some aren't zoonotic (harmful to humans), including E. coli, salmonella and *Clostridium difficile*.

Although you're unlikely to pick up an infection from your dog licking you, it is possible. Older people and those with compromised immune systems are more at risk, and you should be careful with any open wounds you have, as well as licks on your mucous membranes. But our entire world is basically a soup of invisible microbes, so the odd lick probably won't make too much difference.

4.10 Why do dogs sniff each other's bums?

To understand this, we need to change our idea of how dogs see the world. The answer is they *don't* see it – at least not as powerfully as we do. Humans' primary sense is sight, and a large part of the cerebral cortex in our brains is dedicated to processing visual information. **Dogs' brains, on the other hand, are more tuned to processing smell. The area of their brain devoted to smell is 40 times larger than the comparable part of the human brain**. Their 'picture' of the world is built more from smells than sights, something that can be difficult for us to understand.

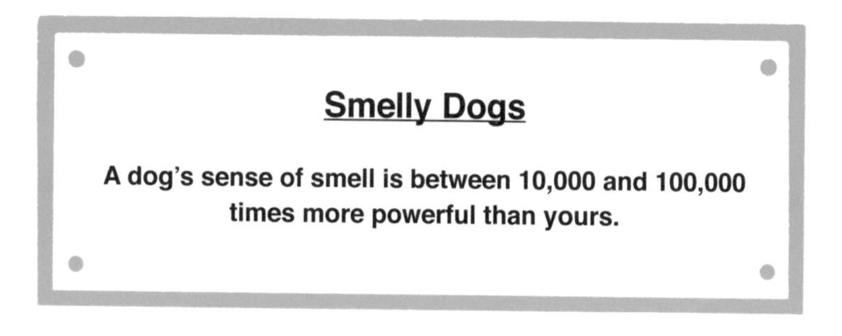

Smelly Dogs
A dog's sense of smell is between 10,000 and 100,000 times more powerful than yours.

Dogs can find out a huge amount of information about other dogs just by smelling their bottoms and genitals (especially the violet gland on dogs' tails, as well as glands beside the anus). Because the vast majority of dogs walk around stark naked, it's all easily available. Bottoms and glands offer traces of information such as age, sex, mood, health, reproductive ability, and stage of reproductive cycle. We can all agree that humans would be much happier if, like dogs, we didn't judge each other on looks alone. Although perhaps the bottom-sniffing thing isn't the best way to kick off positive social change.

Weirdly, female dogs differ from males in their sniffing behaviour. A study published in the journal *Anthrozoös* concluded that 'Female dogs concentrated on [sniffing] the head area, and male dogs the anal area, irrespective of the sex of the other dog'. Funny, that.

Why does your dog sniff *your* crotch? Well, humans give off smells, too, even if we can't sense them. We have a concentration of apocrine sweat glands around our genitals and bottom, and these glands produce pheromones that offer a lot of the same information about ourselves that dogs' do: sex, age, mood, health and stage of menstrual cycle. We may not be the same species, but some of these pheromones are very similar to those of dogs, and males will be especially interested in them, which can lead to dogs becoming sexually excited. Of course, dogs don't understand that we find it rude or unpleasant – they're just interested in the information we have to offer.

4.11 Why do dogs eat poo?

Coprophagia (eating poo) is a revolting concept but surprisingly common in the animal kingdom. I've witnessed baboons eating their own poo and it's been seen in elephants, flies, rhinos, giant pandas and capybaras (the world's largest rodents). Butterflies, flies and beetles all feast on poo, especially from plant-eating herbivores, because it contains lots of semi-digested food. Delish. Termites eat each other's poo to share gut microorganisms that allow them to digest tough cellulose, and lagomorphs like rabbits and hares eat cecotropes (the softer of their two types of poo) to give themselves a second bash at extracting tough plant nutrients. Small mammals such as hamsters and hedgehogs extract nutrients from their own droppings (gut bacteria produce vitamins B and K when they break food down), and some baby animals such as elephants and koalas are born with sterile (bacteria-free) intestines so eat microorganism-rich adult poo to obtain the bacteria they need for digestion.

But dogs don't get any nutritional benefit from poo-eating, do they? No one's entirely sure, but vets have suggested that some dogs eat poo to rebalance their digestive system with bacteria or enzymes they're lacking as a result of a modern diet of over-processed food. **The fact that dogs usually only eat nice bacteria-rich fresh poo rather than crusty old turds** seems to support this. Coprophagia can indicate nutritional deficiency but it's not unusual in perfectly healthy dogs, too, especially puppies. **Coprophagia and rolling in poo are both examples of allelomimetic behaviour – ones dogs learn from watching other dogs**.

Mother dogs lick their puppies poo-ey bottoms and sometimes eat their poo for hygiene purposes, and those puppies may mimic them. So it may just be that a dog's natural curiosity about strong smells has become a habit. The best way to stop your dog eating poo is to gently, firmly and consistently discourage it as soon as you see it happen.

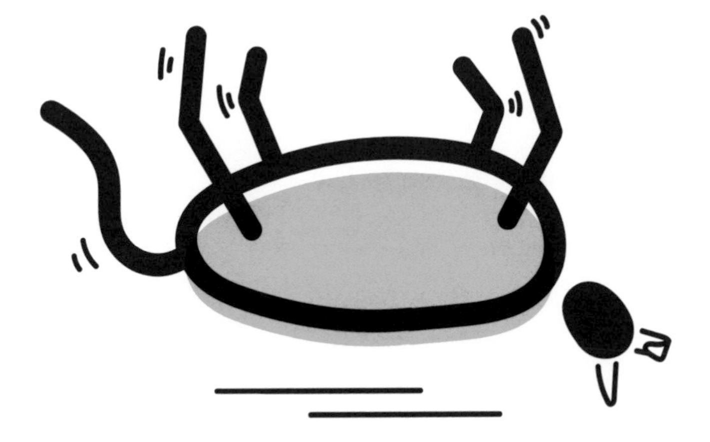

4.12 Why do dogs love lying on their backs with their legs spread?

There's no greater show of self-confidence than an animal lying on its back, legs spread, genitals to the wind, head flung back, jowls drooping the wrong way to expose their teeth: it's the ultimate expression of Living Your Best Life. But it's not just me who does this – dogs do it, too.

When dogs and cats sleep on their backs, they expose their most vulnerable body parts, so this usually happens when they're feeling confident, safe, comfortable and unthreatened. This is perhaps why it's rarely seen in wild animals or dogs sleeping outdoors. **Rolling over and exposing its belly is also a counterintuitive reaction of a submissive dog to a dominant dog**. Although you'd think the worst thing for a submissive animal to do in front of a dominant animal would be to appear vulnerable, it's a way of avoiding a fight and saying 'I don't challenge your dominance'. This version differs from the common sleeping position because it's usually accompanied by other signs of anxiety, including frantic low-angle tail wagging, tail curling and a tense body posture.

There's not much research about dogs lying on their back, but a fair amount of opinion. The more convincing theories are thermoregulation and a good old stretch. Dogs only sweat through their feet, so having all four paws in the air allows for faster evaporative cooling. The fur on their belly is also relatively thin compared to the fur on their back, so exposing it allows heat to dissipate. Lying on their backs also stretches dogs' muscles. In much the same way that we humans enjoy a good stretch, it gives a nice sense of release and relieves joint ache.

Chapter 05: The Very Weird Science of Dog Behaviour

5.01 Do dogs feel guilt?

Almost certainly not. Around three-quarters of owners believe their dogs feel guilt and around half think their dogs feel shame, but it's unlikely they experience either emotion. **Dogs are capable of experiencing primary emotions such as happiness and fear**, and release hormones in response to emotions just as humans do – serotonin and dopamine when happy, and adrenaline and corticotropin when scared. But guilt and spite are seen as more complex secondary emotions that require a theory of mind (the ability to attribute different mental states to themselves and others). Although a few studies show rudimentary evidence of this in dogs that attempt to deceive humans and hide treats from other dogs, it's unlikely the capacity is developed enough for guilt.

So, why do dogs always look so guilty when they've done something that annoys us? When you discover your dog has laid a steaming turd on the carpet or torn apart your favourite shoes, he may well display visual signals that you interpret as guilt: cowering, tail tucked well down, ears turned back, head lowered, eyes looking up at you or avoiding you altogether. But they only *look* like guilt because we prefer to understand things in a human context. To us, it makes sense that an animal we know and love would show guilt after doing something wrong. Deep down, we *want* dogs to look guilty so we can get on with the hard work of forgiving them.

In fact, those signs of 'guilt' are simple indicators that your dog is feeling fear. They are deployed in reaction to your tone of voice and body language (remember that dogs are much better at reading these than humans are). They are also probably learned behaviours:

your dog has remembered from past incidents that if he acts guilty and shows signs of fear, he will be punished less – or your annoyance will stop sooner. An interesting study published in *Behavioural Processes* in 2009 found that **dogs scolded by their owners for eating a snack appeared guilty** ***whether they had done so or not*** (the researchers told some owners that their dog had been naughty by eating a snack even when it wasn't true).

Jealousy is a little different. A study published in 2014 in *PLOS ONE* put dogs in situations where their owners appeared to be showing affection to other dogs or inanimate objects (such as a stuffed dog or book). They found that dogs performed more jealous behaviours – such as snapping, getting between their owner and an object, and physically touching their owner – when another dog was involved, rather than an object. A 2008 study published in the *Proceedings of the National Academy of Sciences* also found that dogs stopped cooperating with researchers when they saw other dogs rewarded with food for actions when they were getting nothing. There may be lots of reasons for this, but it's an indication that dogs feel a primordial form of functional jealousy, and that a sense of fairness is important to them – which makes sense as dogs are descended from social hunters that needed to cooperate and live harmoniously as a pack.

Legendary Dogs

Laika the Space Dog

Warning: this is not a happy story.

Stray dogs in Moscow have to withstand extreme cold and hunger, which is why in November 1957 Russian scientists chose a calm (but barky) stray mongrel first named Kudryavka ('Little Curly') and then Laika ('Barker') to become the first animal to orbit the Earth. However, the trip in Sputnik 2 was always doomed to end badly as there was no plan to bring Laika back alive – a fact that was announced shortly after launch, to the outrage of some observers. The Soviet government initially claimed that she had been euthanized before running out of oxygen, but in 2002 it was revealed that her pulse rate had tripled and her breathing rate had quadrupled during launch. Despite reaching orbit alive, she probably died within five to seven hours from overheating and stress. The Soviet Union is thought to have launched dogs into space on 71 flights between 1951 and 1966, and 17 of them died. In 1997 a statue in memory of both Laika and human cosmonauts was unveiled in Russia's Star City.

5.02 Why do dogs wag their tails?

Sounds obvious, doesn't it? Dog wags tail, therefore dog is happy. But it's way more interesting than that. The clue is in the fact that puppies are entirely capable of wagging their tails from birth, but don't do so until they're about six or seven weeks old, when they start socially interacting with each other.

It's thought that tails originally evolved to help with balance: dogs swing their tail from one side to the other when walking along narrow surfaces to correct any body tilt. Tails also help dogs to make sudden turns when running at high speed, working as a counterbalance to stop them spinning out of control – which would have been especially useful when hunting.

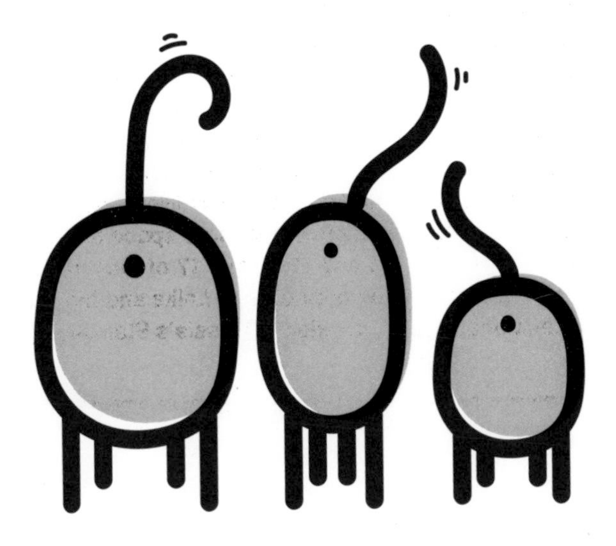

But at rest tails are of little physical importance, so evolution stepped in to co-opt them as a communication tool. Dogs are social animals living together in packs (unlike cats), so it's hugely important that they have multiple communication signals to help fend off attackers, and to hunt, live, breed and raise their young together with as little conflict as possible. As mentioned previously, puppies first begin to wag their tails when they start to interact with each other. They often wag during feeding to indicate that they come in peace, even if they have been play-fighting moments earlier.

Tail wagging can indeed signal happiness, but it can also mean fear, a challenge or worse – it's part of a combination of signals that dogs are better at reading than humans. The height of a dog's tail wag in relation to its usual height (breeds can have very different resting tail heights) is highly expressive: **mid-height wagging indicates that a dog is relaxed and a low wag indicates submission, but a vertical tail is a signal of domination and if it's wagging in small, fast movements, it's possibly a sign of impending attack.**

The Longest Tail

The longest dog tail ever recorded measured 76.7cm (30.2 inches) and belonged to Keon, an Irish Setter from Westerlo in Belgium, according to *Guinness World Records*.

5.03 What does left- or right-handed tail wagging mean?

Dogs wag their tails more to the left or the right depending on their emotions. In a study published in *Current Biology*, 30 different pet dogs were each shown four items: their owner, an unfamiliar human, a cat and an unfamiliar, dominant dog. When they saw their owners, their wag was vigorous and biased more to the right, whereas the unfamiliar human made them wag only moderately and to the right. The cat made them wag slowly and with restraint to the right, but the unfamiliar, aggressive dog made them wag to the left.

It seems that **when dogs feel positive about something, they wag their tails more to the right, and when they feel negative, they wag more to the left**. This bias may simply be an evolved directional signal to help dogs communicate: right-wagging tails have been shown to relax other dogs passing by, while left-wagging ones seem to stress them out.

But it has also been suggested that it's a function of dogs' brains having differences between left- and right-hemisphere control: when the tail wags to the left, it is mainly under the control of the right hemisphere of the brain, which tends to govern and express intense emotions such as fear and aggression, and hence withdrawal (negative responses). When it wags to the right, it's under the control of the left hemisphere, which governs more positive emotions.

Wagging is Complicated

Wagging tails are context specific, so be careful. It could mean that a dog is happy or curious, but it can also mean it is about to attack.

Other studies have shown that dogs tend to turn their heads to the left in response to threatening stimuli, and many animals, including toads and horses, show stronger avoidance responses when they see a potential threat on their left rather than on their right. The same goes for dogs' reactions to sound: research found that the scary sound of a thunderstorm made dogs turn their heads to the left, while more familiar dog barks made them turn to the right.

5.04 How clever is your dog?

A huge amount of scientific research has been carried out on dog behaviour for the simple reason that dogs, in contrast to cats, are spectacularly easy to study. They are biddable, food-reward orientated, they enjoy pleasing humans, and are highly adaptable to research environments. You can even get them to work to command in notoriously noisy MRI scanners to allow scientists to study their brains. Try doing that with a cat and you're likely to lose an eye. But despite their reputation for being highly intelligent, dogs have a relatively small brain – about the size of a lemon depending on the breed – and don't perform as well as other animals in many tasks.

Most animal cognition researchers say that comparing the cleverness of dogs and humans (or of whales or ants, for that matter) is irrelevant. Dogs are as clever as they need to be to ensure the survival of their species: specifically, a recently domesticated canine, predatory, social hunter. In the same way, whales' intelligence and physical ability are specific to their underwater environment, food sources and predators, while ants live in complex societies that simply wouldn't function properly if they had the same intelligence, needs and individuality as humans.

Brain size in itself isn't a particularly good indicator of intelligence, blue whales' brains weigh up to 9kg (20lb), while desert ants' are 0.00028g (0.00001oz). **But brain size *relative* to body size does make a difference, and in this, dogs do pretty well with a ratio of 1:125 compared to humans' 1:50 and horses' 1:600**. That said, we can still make a few useful comparisons in areas such as memory, self-awareness and numerical, sensory, spatial, social and physical cognition (understanding of the physical world).

It's thought dogs can remember around 165 words and commands (the brightest 20% can learn up to 250 words), count up to four or five, and deceive humans and other dogs in order to get rewards. The latter is particularly important as it means they may have a rudimentary theory of mind (which, as we've mentioned, is an ability to attribute various mental states to themselves and others). Dogs are extraordinarily good at reading human gestures and expressions and can follow a human's pointing (cats, elephants, seals,

Legendary Dogs

Lassie

In 1940 the English writer Eric Knight published the novel *Lassie Come-Home*, which became an MGM hit movie in 1943 (tragically, the same year that Knight died in an air crash in South America). Lassie was renowned for running to fetch help, leading people away from danger and taking strays back home. She became a regular and much-loved character in many movies and TV series, including the CBS show *Lassie*, which ran for an astonishing 591 episodes between 1954 and 1973, with the title role played by numerous different Rough Collies. The first Lassie was a male called Pal (though the fictional Lassie was female), who appeared in the first six films and two pilots for the TV series. Many of his descendants went on to play the character in the movies and episodes that followed.

ferrets and horses can also do this to some extent). This is significant because it's a cross-species sharing of conceptual information (the dog understands that we are interested in something that we want to share), which is probably why so much scientific research has been carried out on it. **Pointer breeds also return the favour by freezing, often with one paw raised and muzzle facing the direction they want us to look in, to help us spot something interesting, such as prey.**

One fascinating aspect of canine sophistication is intelligent disobedience. This is the extraordinary ability that guide dogs have to refuse their owner's command if they perceive it would put them in danger – thus contradicting dogs' learned behaviour to obey their owners. But it gets even more complex: if the dog refuses to move because, say, there's a flight of stairs ahead, the owner can override the dog's objection using a code word that shows they understand the stairs are there. If the owner uses the wrong code word, however – maybe because they think they're facing a kerb rather than stairs – the dog will still refuse to move. Only when the dog hears the correct code for the situation will it let the owner proceed. And if the dog perceives a danger such as a cliff or precipice, it will simply refuse to move forward. A Yale study has shown that **dogs are better than three-to-four-year-old children at ignoring bad instructions**.

Female dogs – but not males, interestingly – also understand object permanence. This is the idea that objects shouldn't change into something else and that, when out of sight, they haven't necessarily vaporised into non-existence – something that humans take a long time to learn. To some degree, dogs also show a sense of time by anticipating when their owners are likely to return home.

Dogs can recognize both dog and human emotions, they use humans to solve some problems for them (which could imply either intelligence or the lack of it!), and have episodic-like memory (memory of daily events). **Rico, a Border Collie studied by the Max Planck Institute for Evolutionary Anthropology in Leipzig, knew the names of 200 items and was able to fast-map (form a quick hypothesis about the meaning of a new word) by inferring the name of a new item by exclusion**. That's to say he worked out that when he was shown a new item he hadn't seen before, and then asked for a new item with a name he hadn't heard before, both new name and new item must correspond. He could also give items to specific people. This might not sound much to you, but it's a task humans only achieve by around the age of three.

On the downside, a 2018 study published in *Learning and Behavior* concluded that 'dog cognition does not look exceptional' in comparison to chimps, dolphins, horses and pigeons. Dogs' perception and sensory cognition is excellent, although similar to that of many other species; their spatial cognition is also good but not exceptional, and their physical cognition is lacklustre compared to some other species'. Dogs' social cognition is excellent but chimps are more likely to exhibit deception and empathy, and both chimps and dolphins do far better in self-consciousness tests. Pigeons' pattern recognition and homing capacity far exceeds that of dogs, chimps can use tools, raccoons do better at string-pulling tasks, and sheep may well be better at facial recognition. But dog lovers shouldn't be disheartened: although dogs don't necessarily outperform other animals in specific areas, they perform consistently well across lots of different intelligence categories, and are brilliantly clever at being exactly what they are: recently domesticated, predatory social hunters. Isn't that enough?

5.05 Does your dog love you (or just need you)?

Of course we all *think* our dog loves us – my dog is always happy to see me, hangs out with me, craves playing with me, wags his tail and licks my face in greeting. But does he just do these things to get what he wants: attention, a mouthful of croissant, or to play with a ball? To get to the bottom of this, we have to set aside our soft, fuzzy concept of love as an undefinable, magical gooiness and strip emotion down to its nuts and bolts (with apologies to all the romantics out there), where love is all about biochemistry.

We can break love down into three basic biochemical systems, marked by different hormones (chemicals produced by glands in the body that affect activity and behaviour). The first system (which doesn't really apply to love between humans and dogs, but is interesting nonetheless) is sexual attraction, whereby the hormones testosterone and oestrogen are released, both of which are a vital part of each species' reproductive cycle.

The second system is infinitely more important to loving your pet: attraction. The chemicals associated with it are dopamine (a hormone and neurotransmitter that allows us to feel pleasure), serotonin (a complex neurotransmitter that many call 'the happiness chemical' but might be better described as a mood stabilizer and, more weirdly, a blood-clotting agent), and norepinephrine (a hormone that increases excitement and arousal). When we interact with dogs, levels of these hormones change for both them and us.

The third biochemical love system (good name for a band) is attachment, and again there's more biochemical crossover. When

owners are in contact with their dogs, their hormone levels mirror those for human attachment – oxytocin levels multiply by five, and endorphin and dopamine levels double (interestingly, this also happens when dogs and owners stare into each other's eyes). These levels are associated with bonding, pleasure and joy.

There's other evidence of dog love. Some amazing research by neuroscientist Dr Gregory Berns used fMRI scans (which measure brain activity by detecting changes in blood flow) to show that only familiar humans activated activity in dogs' caudate nucleus (part of the brain associated with positive expectations), whereas strangers did not.

The problem is that **although we know that dogs share the same basic biochemistry of love with humans, they aren't able to tell us if their *experience* of that love is the same as ours**. But the fact that this human-to-dog physiological response is often stronger than dog-to-dog response is remarkable. Of course there's also the tantalizing possibility that dogs may feel love, happiness or joy *more* strongly than humans do, rather than less, which allows a bit of psychological wriggle room for anyone who feels that all this biochemistry has taken the romance out of the whole shebang.

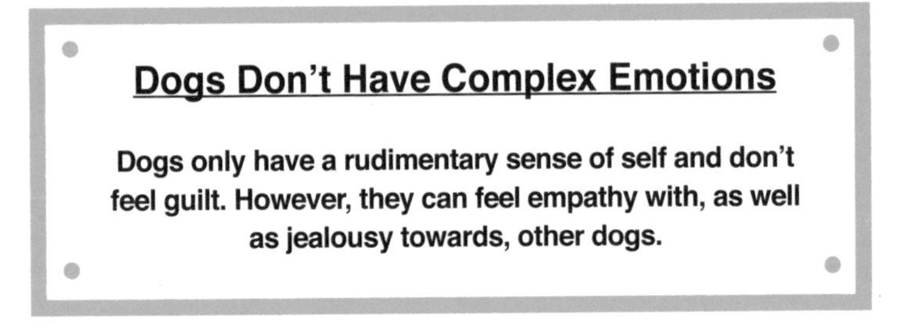

Dogs Don't Have Complex Emotions

Dogs only have a rudimentary sense of self and don't feel guilt. However, they can feel empathy with, as well as jealousy towards, other dogs.

5.06 What is my dog thinking?

It's hard to know how dogs think because they are rubbish at telling us. That said, it can be hard to know what our fellow humans are thinking, even though we have complex communication tools such as language, art, music, drama and dance to help us. One person may feel love, guilt or faith in a different way to another, just as a dog may feel joy, but in a different form to us.

Clever Dogs
Dogs are about as intelligent as a two-year-old child. The average dog can understand around 165 words and gestures.

But we can take a look under the hood to make some sense of a dog's thoughts. In an attempt to understand the canine mind, neuroscientist Dr Gregory Berns and his research team trained dogs to voluntarily hop into fMRI scanners so their brains could be scanned, explaining his findings in the fantastic book *What It's Like to Be a Dog*. He found that there are startling similarities in how dog and human brains function (and those of all mammals). Berns deduced that, because we have many neural processes in common, we are likely to have similar subjective experiences. He also concluded that because different dogs appear to have different neural responses to the same experience, it's likely that in neurological terms dogs do have

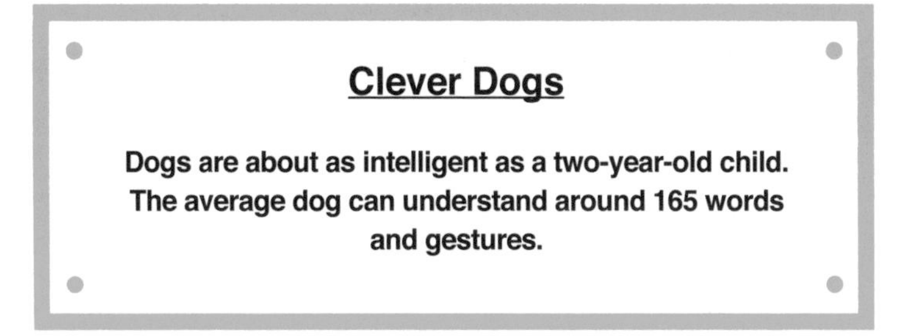

individual personalities. Dog owners might think this is obvious, but it's something animal behaviourists have been wary of confirming.

So, do dogs feel empathy for humans? Can they share our feelings? A 2011 study published in *Animal Cognition* found that when dogs were presented with a crying stranger, they sniffed, nuzzled and licked the stranger rather than their owner. The researchers concluded that dogs express empathic behaviour towards humans. But they added that, in strict scientific terms, this might also be interpreted as 'emotional contagion' coupled with learned behaviour – rather than showing true empathy, they could just be reflecting the fact that they had 'been rewarded for approaching distressed human companions' in the past.

Dogs do feel a strong affection towards us; some research even shows that they are more interested in people than other dogs. It's easy to take this for granted because we forget how unusual it is for one species to play with and relate to another. A 2015 study published in *Current Biology* used fMRI scans of dogs' brains to show that they could recognize the difference between happy, sad, angry or fearful facial expressions made by unfamiliar humans.

On its own, none of this proves that dogs think like us, but there are intriguing hints. There's evidence that dogs prefer people they see helping their owners, that their brains respond to laughter and barking in the same way as humans, and that they share emotions with us. But my favourite piece of canine empathy research comes from Biagio D'Aniello of the University of Naples Federico II in Italy, who found that **dogs can *smell* our emotional state from our sweat, and that they then *share the same emotions*.** I know!

5.07 Why do dogs yawn?

No one's really sure why humans yawn, but we know it's related to tiredness and boredom – and also stress – and that it's contagious. At school we used to have secret yawning competitions where we would all yawn as ostentatiously as possible until we managed to make the teacher yawn, too. Happy days. It's different for dogs, though.

Dogs do sometimes yawn from tiredness, but more often it's down to stress and anxiety. My dog Blue yawns over and over again when he knows we're about to go for a walk but I'm frustrating him by making coffee, lacing my boots and endlessly nipping back inside because I've forgotten some essential dog-walking tool (travel mug, book, tennis balls (x 2), headphones, poo bags, keys, brain …). Each yawn has the required effect of making me hurry up, getting more flustered and forgetting more things. Poor fella.

Dog trainers report that dogs not doing well in training yawn frequently, while professional dog walkers say yawning is often a passive dog's response to an aggressive dog. **Yawns are also contagious within packs of wolves and within packs of dogs, especially when they are stressed, but the best yawny fact is that non-stress yawns are contagious between humans and dogs, too**. Researchers at the University of Tokyo found that a dog is more likely to respond to a human yawn with a yawn of their own if the yawner is familiar to them. They concluded that it was an expression of empathy: 'Contagious yawning in dogs is emotionally connected in a way similar to humans.'

5.08 Why do dogs tilt their heads when confused?

Many dogs tilt their head to one side when their owners talk to them. My dog does this when he doesn't understand what I'm saying, often because I'm deliberately talking gobbledegook to him just so he'll tilt his head – it's so cute!

There's little research and no clear answer as to why dogs do this, which leaves us in the fruitcake-heavy badlands of *opinion*. So here's a potted rundown of some of the opinions out there.

Many vets think that dogs tilt their heads when confused because it's a physical movement related to other auditory problem-solving – specifically, when they're trying to locate the source of a sound. This is a relatively convincing theory as dogs do seem to tilt their heads in response to words they don't understand. The drawback is that head-tilting doesn't make a blind bit of difference to dogs' understanding of what you're saying. That said, dogs have many evolutionary quirks that are useless, but not negative enough to have been bred out of them.

Psychologist Stanley Coren has written many books on dog behaviour and cognition, and thinks that dogs' snouts obstruct their vision, so tilting their heads allows them to get a better view of our faces, and especially our mouths. On the other hand, Steven R Lindsay's *Handbook of Applied Dog Behaviour and Training* deduces that **the muscles of dogs' middle ears are controlled by the same part of the brain that's responsible for facial expressions and head movements**, so when they tilt their head, they're trying to work out what you're saying, and communicate to you that they're listening.

Possibly more convincing is the simpler theory that it's a learned behaviour – maybe dogs simply enjoy our positive response when we see them tilt their heads.

5.09 What the heck are the 'zoomies'?

Every now and then, dogs seem to just go nuts, running from room to room, zipping up and down the furniture with legs spinning like Wile E Coyote, sometimes chasing their tails or running in circles. My dog's 'zoomies' are triggered by having a shower or bath, and he seems to thoroughly enjoy them. **These strange outbursts of energy are pseudo-scientifically known as FRAPs (Frenetic Random Activity Periods)** and little is understood about them, although they're very common and have been reported in cats, too.

In the absence of any hard data about the zoomies, we're at the mercy of newspapers, magazines, bloggers and other assorted people with *opinions*. Now, opinions may be lovely, heartfelt, and one or two may *even* be correct, but they are no substitute for facts. Nonetheless, here are a few:

1. The zoomies don't seem to be associated with any neurological problem, and may even be beneficial to the dog. As long as they don't run headfirst into the dishwasher.

2. Don't chase a dog with the zoomies as they get over-excited and less dexterous, and may end up running headfirst into the dishwasher.

3. They are most often noticed after dogs have eaten, been washed or come back from a walk, as well as before going to bed.

4. They are more common in puppies and younger dogs.

5. The reason no one has done any research into them is because they don't seem to cause any problems for dogs or owners and therefore aren't worth the money or time to study.

6. Other names for the zoomies include 'puppy demons', 'hucklebutting' and 'frapping'. These are all wrong. They're called the zoomies. End of.

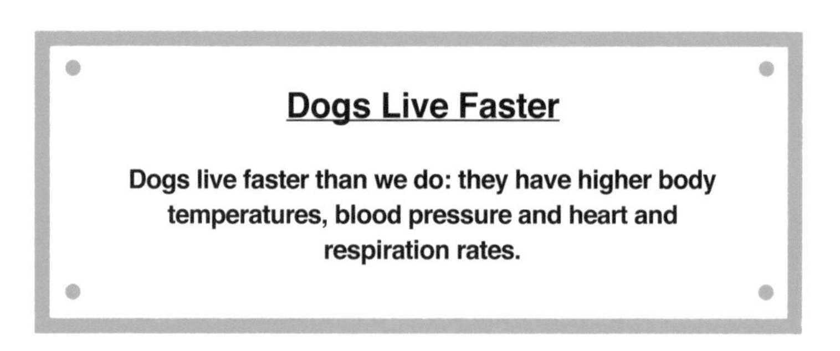

Dogs Live Faster
Dogs live faster than we do: they have higher body temperatures, blood pressure and heart and respiration rates.

5.10 Do dogs dream? If so, what about?

Although unencumbered by either gainful employment or addiction to Minecraft, dogs probably don't sleep as much as you'd imagine. One study showed that Pointer dogs spent 44% of their time alert, 21% drowsy, 12% in REM (Rapid Eye Movement) sleep and 23% in deep slow-wave sleep.

It's impossible to know for certain if dogs dream because they're rubbish at telling us, but all the neurological evidence suggests they do. **Their brains show wave patterns and activity that are similar to humans' when asleep and they go through similar sleep stages, including REM**, complete with irregular breathing patterns and flickering eyelids. This is when they are most likely to be dreaming – when humans are awakened during REM sleep they will often say they were in the middle of a dream. My dog Blue will sometimes grumble, growl and whimper at the same time (I'll call out to calm him, but whether that makes any difference is anyone's guess) and his legs will twitch, leading me to think he might be dreaming about chasing squirrels. He does love chasing squirrels.

What do dogs dream about? If we make a comparison with rat – and human – dreams, it's likely they are remembering things that happened during the day, or re-enacting their usual activities: going for walks, defending the house, chasing squirrels, running, stealing balls, chasing squirrels, herding the family, chasing squirrels, chasing squirrels, chasing squirrels and chasing squirrels.

5.11 What was Pavlov on about?

People sometimes use the phrase 'Pavlovian response' to describe how a sound can be paired with the serving of food in order to teach a dog to salivate in response to the sound itself, rather than just the food. This is called a conditioned reflex and it's all rooted in classical conditioning. Ivan Petrovich Pavlov, a Russian psychologist who died in 1936, was its leading authority, having noticed almost by chance that dogs salivated at different times and rates while he was researching canine digestion.

Pavlov set up an experiment in which a buzzer or metronome sounded at the same time as food was given to a dog, making the dog salivate. Through classical conditioning, the dog associated the sound with being fed, and would salivate whenever it heard it. From this, Pavlov developed a whole set of behavioural theories now used in lots of different situations, especially by teachers in classrooms. Teachers will often use classical conditioning to manipulate the classroom environment to help positive learning or induce comfort (and sometimes fear) in their students – perhaps by dimming the lights, holding up a detention list, or doing an activity such as clapping three times, which students have learned to use as a cue to be quiet.

Incidentally, most people think Pavlov used a bell to signal the arrival of food, but there's no evidence he ever did. Instead, he used buzzers, metronomes and sometimes electric shocks. If you want to appear knowledgeable, you should relate this fact whenever anyone mentions Pavlov – although this is unlikely to make anyone like you.

5.12 Why do dogs bury things?

Burying food is known as 'caching' and it's an evolutionary remnant of a hard-wired natural survival instinct. Dogs' ancestors would have been more likely to survive if they had an instinct to hide excess food from predators and the rest of the pack, which they could then dig up later when hungry. Of course, domestic pets don't need to be so protective of their food now their owners lovingly cater for their every need, but evolution can take a long time to adapt, even when a behaviour is no longer relevant. The fact that dogs also pointlessly stuff food down the back of the sofa and bury toys reveals the strength of this instinct beyond its rational use, although too much caching is sometimes diagnosed as problem behaviour in bored, anxious or defensive dogs.

The selfishness of the caching instinct may seem odd for social animals that evolved hunting together, but even a successful wolf pack can have a tricky time dividing up the spoils of a hunt, and fights sometimes break out – although parents and siblings freely share their food with puppies.

Dogs share this burying instinct with squirrels, hamsters, many birds and also humans. I stayed with an Inuit family in the Northern Canadian Arctic Circle who buried walrus carcasses in the ground for months on end, which would make the meat ferment and mature. They dug a huge chunk up for me to try and although I found the flavour gruesomely pungent, the whole family loved it, including a tiny tottering two-year-old.

5.13 Why do dogs love to play?

It sounds obvious, doesn't it? Dogs play because it's fun! But in behavioural science terms, fun alone isn't a good enough reason. Play takes time and effort, and evolution dictates any activity that distracts from hunting, eating or reproducing must help a wild animal's likelihood of survival – otherwise it would have been bred out of them. Many young mammals play (wolf puppies will even play fetch with humans), but dogs are unusual in that they have a huge appetite for play even as adults. It's likely that artificial selection is at work here – when dogs were domesticating we selected those that displayed the most puppy-like behaviour because we thought it was attractive.

There's lots of evidence that play teaches animals social skills and tests and strengthens social bonds (which is especially important for pack animals). It also helps their physical and cognitive development, builds up their emotional flexibility for dealing with the unexpected and helps them understand their abilities in relation to other animals. But there's little evidence surrounding the mechanisms involved in animal play and, after more than a century of research, little scientific consensus on its evolutionary function. Why does social play make a species any more successful?

Many bouts of play tread a fine line between friendliness and aggression: play-biting, mounting, chasing and pinning each other down. To ensure peaceful play (and to encourage it to continue), dogs use lots of different signals to invite each other to play, and to synchronize their activities. **The classic invite to play is the 'play bow' where the dog lays her front legs on the floor and raises her behind in the air**, often while barking and

wagging her tail. This will also happen during play itself, with a brief pause followed by another play bow to initiate play once again. Other signals are a bounding run, high-pitched barking, bowed head, clawing action and sometimes a mock-retreat move.

What we can be sure of is that play generates happiness: a release of hormones that simply make a dog feel good. Artificial selection has its pitfalls (some physical breeding characteristics encouraged by humans have been harmful for dogs), but the fact that we have selected for playful, happy dogs, and that happy dogs tend to be well-socialized ones *ought* to be good for the species.

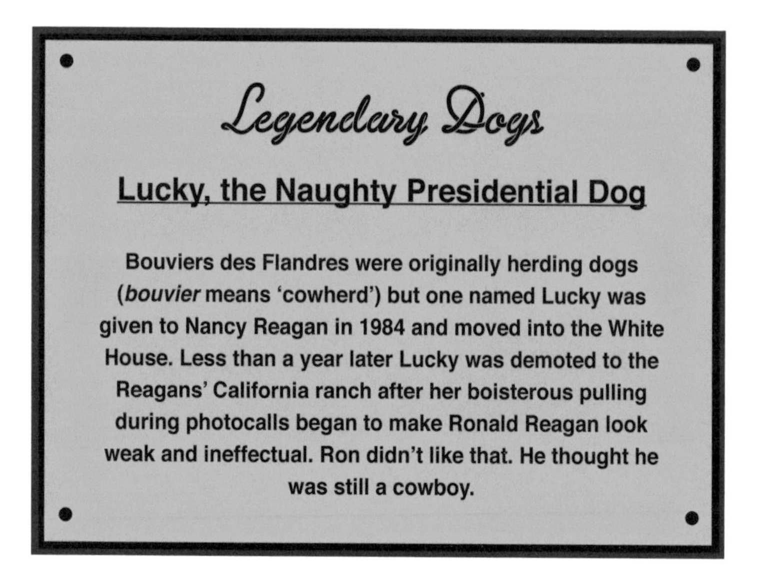

Legendary Dogs

Lucky, the Naughty Presidential Dog

Bouviers des Flandres were originally herding dogs (*bouvier* means 'cowherd') but one named Lucky was given to Nancy Reagan in 1984 and moved into the White House. Less than a year later Lucky was demoted to the Reagans' California ranch after her boisterous pulling during photocalls began to make Ronald Reagan look weak and ineffectual. Ron didn't like that. He thought he was still a cowboy.

5.14 Why do dogs love eating shoes?

Dogs are genetically predisposed to enjoying a good chew. This is possibly because those doggy ancestors who loved chewing bones could access extra calories from the marrow inside. Individuals predisposed to chewing – whether bones or shoes – were thus more likely to survive and pass on their genes when food was scarce. Dogs inherited this habit, which is no longer relevant in their lives as pets.

Chewing is annoying, but it's something you as a pet owner may just have to deal with. Although we can train dogs to adapt some of their behaviour, we can't get them to adjust everything to fit in with our lives without them losing the essence of dog that makes us love them in the first place.

Many young dogs chew to relieve the pain of teething, and may return to it when older for several reasons: 1) it's a good activity to combat boredom and frustration; 2) it relieves separation anxiety; and 3) it may also be related to simple hunger. But why *shoes*? Well, there are several good opinions (but no real research) as to why they pick on your favourite footwear:

1. At the simplest level, shoes fit into a dog's mouth – in fact they're often the size of a good bone.

2. Your shoes smell of you (for better or worse) so your dog will naturally be interested in them.

3. Shoes are made of materials that are eminently chewable, such as leather, rubber and canvas – soft and pliable but resilient, they also break down with a bit of perseverance. This makes them a good challenge – you don't get the same from chewing a laptop. Thank God.

The Fastest-Cycling Dog

Norman the Briard holds the Guinness World Record for the fastest 30m (100ft) cycled by a dog – 55.41 seconds. He had stabilizers, but it's still quite a feat.

5.15 Can dogs really find their way home from miles away?

Dogs' homing ability is the stuff of legend, immortalized in the classic Walt Disney tearjerker *The Incredible Journey*, in which two dogs and a cat lose their owners while on holiday and travel miles in search of them. But it's also a real, tangible and well-documented phenomenon. In 1924 Bobbie the Wonder Dog travelled 4,500km (2,800 miles) from Indiana back to his home town in Oregon, and stories of shorter trips abound: 92km (57 miles) to return to an old house, 18km (11 miles) to return to a foster home, and a 29km (18-mile) trip that included traversing a wide river.

Of course, if we look at this dispassionately, we could assume it's random luck – the number of dogs lost every year far outnumbers those that find their way home, and after all, newspapers will always report the rare dog that finds its way back from some distance away but never once mention the thousands of lost dogs who don't return. However, there is some science to back the idea up.

Scent plays an important part in the homing process, and you have to remember that dogs' sense of smell is powerful to an extent we can barely imagine. It's possible that dogs can retrace their steps for miles, and even if they have been driven somewhere and not left a scent, they may be able to track familiar scents of other dogs, fields, restaurants or farms to eventually bring them home.

But it gets much more fascinating: an extraordinary 2020 study published in the journal *eLife* looked at hunting dogs over three years and found that **30% of them scouted their way back to their owners using magnetic navigation. They started**

with a short 20m (22-yard) 'compass run' along a north–south axis to get their geomagnetic bearings, and then successfully headed back home using routes that made no use of scent. This geolocation ability may sound far-fetched but it ties into the fact that dogs prefer to poo along a north–south axis (see p28).

Legendary Dogs

Kuno, the World's Bravest Dog

The Dickin Medal for animal gallantry in conflict was first awarded in 1943 to recognize bravery in World War II, and was then revived for animals in all military conflicts in 2000. Well over half the original recipients were pigeons, but since 2000 the vast majority have been dogs, including the extraordinary Kuno, a Belgian Malinois belonging to the British Army, who was decorated in August 2020 for courage during a Special Boat Service raid in Afghanistan. Wearing night-vision goggles, he attacked a gunman and wrestled him to the ground despite having been shot in both hind legs. He survived, but on his return to Britain one of his hind paws had to be amputated and he had a prosthetic fitted.

5.16 Why do dogs chase their tails?

Some dogs just love chasing their tails, much to the amusement of humans. And it's possible that the fact that we show them this amusement is one of the reasons they do it. It's positive reinforcement: if your dog notices that the first few times she chases her tail you give her positive, loving attention, she's likely to do it again. It's almost a reverse training scenario – she gets pleasure from your attention, so she's trained *you* to show her that attention.

But why do dogs start chasing their tails in the first place? Their fast visual flicker-fusion rate developed to help them hunt quick-moving prey (see p97), so they get excited by fast-moving objects, whether it's their tail or the long-suffering cat in your house. If they catch sight of the end of their tail flicking as they turn, it can trigger that obsessive spinning that amuses us so much.

If tail chasing becomes a regular thing, it might be a sign of a problem such as canine compulsive disorder, which seems to be related to vitamin and mineral deficiency. Obsessive tail-chasing dogs tend to be shyer than most, have often been separated from their mothers early, and many will show other compulsive behaviours. Tail chasing can also be a reaction to fleas or ticks, injury, or sheer boredom. But if you give your dog lots of good attention and exercise, and she just chases her tail occasionally, enjoy.

5.17 Why do dogs walk in circles before lying down?

Many animals – including humans – spend time preparing their beds before settling down to sleep. Dogs are no different, often circling around and scratching at bedding before lying down. It's a distant echo of behaviour that was useful to dogs' wolf ancestors, who would have needed to check the ground for pests such as insects and snakes, and to trample down snow, leaves or sharp vegetation. They may also have flattened an area to show the rest of the pack that it had been claimed, and spent time choosing a position that gave the best balance of comfort, warmth and safety from predators. **Circling also allows a wolf to get a good sense of the rest of the pack's whereabouts, especially that of youngsters in need of protection.**

Even though your dog may own a fancy modern fur-lined bed that's set you back a week's wages and there's not a blade of grass in sight, she still has a residual wolfiness that draws her back to hard-wired ancient behaviours that are no longer useful. Your dog has many of these redundant quirks for the simple fact that there's been no strong enough reason for them to have been bred out – known as 'selecting against habit'.

Chapter 06: Dog Senses

6.01 Dog smell

B

Dogs' faces are dominated by one enormous bloomin' great schnoz, and for good reason. With between 125 million and 300 million scent receptors compared to humans' 5 million, and a 40-times-larger brain area dedicated to interpreting smell messages, dogs' sense of smell is their undoubted super-skill. Their smell perception is 10,000–100,000 times more accurate than ours, they can inhale up to 300 times a minute in short sniffs, and **their noses can detect some substances at concentrations of one part per billion (the equivalent of one teaspoon in two Olympic-sized swimming pools)**. Their nostrils can even swivel to let them work out which direction a smell is coming from.

Sniffing is a disruption of ordinary breathing using tiny, short breaths that allow the smell molecules in the nose to stay in place for longer to allow for detection – if dogs breathe normally in long breaths while searching for smells, those molecules are more likely to be dislodged. As dogs inhale, air passes over a labyrinth of scrolled-up plates called turbinates, which are lined with smell receptors covering an area of 18–150cm^2 (3–23 sq in), compared with humans' 3–4cm^2 (0.5–0.6 sq in). They also have something called a vomeronasal organ at the bottom of the nasal passage – it's like an extra nose tuned to identifying pheromones containing signals useful for mating and socializing.

Why are dogs' noses wet? The tip of a dog's nose, known as the rhinarium, stays moist so that thermoreceptors on it can detect wind direction by evaporative cooling (the side that is coldest will be where the wind is coming from), and this can help with navigation and

locating the source of smells and sounds. Recent research published in *Scientific Reports* has even hinted that the rhinarium is used to perceive faint sources of infra-red heat, too. There is some controversy over whether it has smell receptors of its own, or whether its main use is to redirect pheromone smells to the vomeronasal organ by changing its shape.

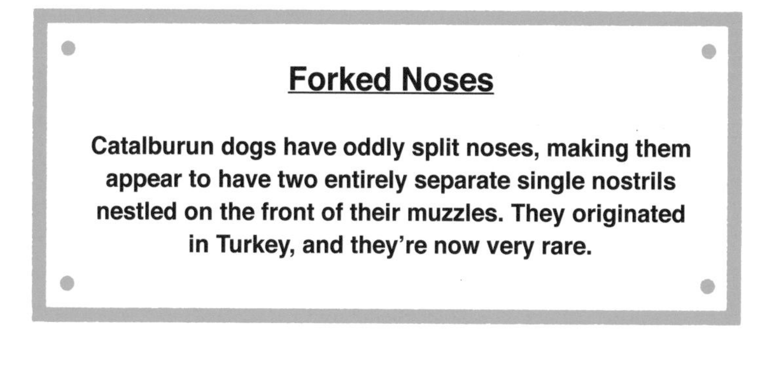

Forked Noses

Catalburun dogs have oddly split noses, making them appear to have two entirely separate single nostrils nestled on the front of their muzzles. They originated in Turkey, and they're now very rare.

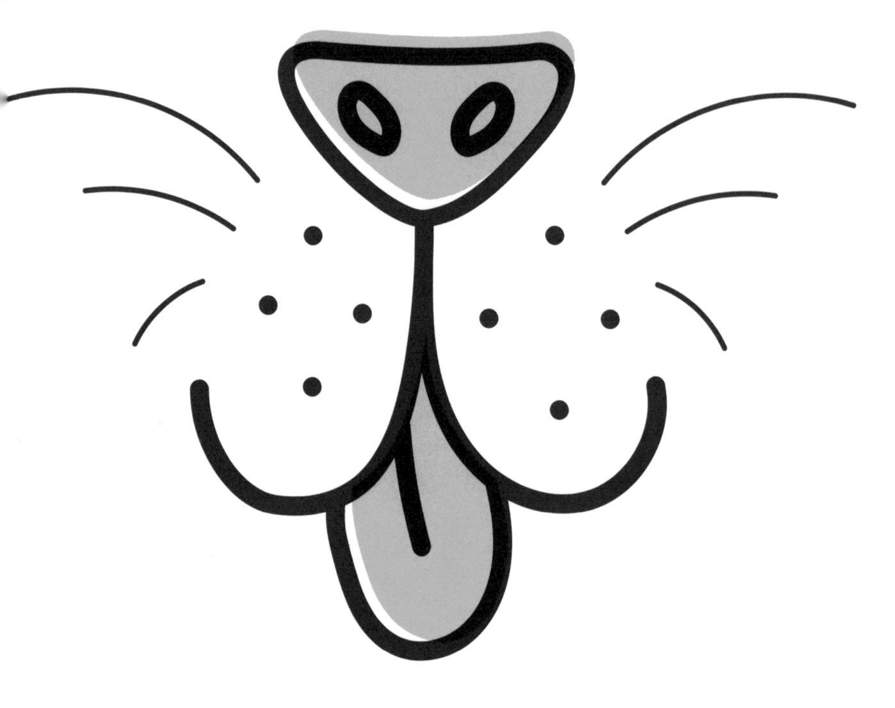

6.02 Can dogs really diagnose disease?

Humans have made use of dogs' phenomenal sense of smell for thousands of years, putting them to work as trackers, hunters and intruder alarms. But the most extraordinary development in the human-dog relationship has been the use of dogs as biodetectors, identifying diseases or conditions by smell alone. It may sound bizarre, but we have to remember that dogs' sense of smell is powerful on a level that we can barely imagine.

From a dog's point of view, each human is enveloped in a cloud of very specific odours, the blend of which creates our personal signature. Many of these come from the environment, but they also ooze out of us in sweat, breath, mucus, urine, faeces and the unique blend of bacteria, yeasts and fungi that live on our bodies. That **human odour cloud can change when diseases and conditions cause chemical alterations in our cells that are expressed in the production of volatile organic compounds**

Dog Diagnosis

Dogs can be trained to use their powerful sense of smell to detect cancer, diabetes, and even epileptic seizures *before* they happen.

(VOCs – basically, active molecules that have a smell). Cancer VOCs were first identified in 1971 in human urine, and they also make their way into our breath and sweat to create a signature smell called a biomarker, which dogs can be trained to recognize with surprising accuracy.

Training dogs to recognize biomarkers is a long and expensive process, though some diseases and conditions seem to be easier for them to identify than others. They are remarkably effective at spotting prostate cancer (which is notoriously hard to accurately diagnose), as well as identifying epilepsy seizures (even before they happen), skin, lung, bladder and breast cancer, and malaria (from smelling children's socks). They have also been put to work detecting Covid-19.

One particularly exciting development is that **dogs can detect Parkinson's disease *before* symptoms start to show. This could be groundbreaking** as symptoms usually begin when more than half of the associated nerve cells in the brain have already been lost. Early diagnosis could be life-changing.

The *Lassie* idea that dogs instinctively understand when we're in trouble and will seek help isn't, sadly, borne out in research conducted by William Roberts of Ontario's Western University. In the study, dog owners walked to the middle of a field with their dogs, then fell to the ground feigning a heart attack, while two other humans sat nearby reading. Although the owners lay still for six minutes, not a single dog sought help for their owners. This either means that dogs are great at spotting bad acting or that *Lassie* wasn't a documentary after all. Was my *whole* childhood a lie?

6.03 Dog sight

Sight isn't dogs' super-skill and their world isn't dominated by vision in the way that ours is, but it would be wrong to think their eyes are worse than ours – they're just *different*. **Dogs are colour-blind. They do see some colours, but their eyes only have a two-colour dichromatic colour range of blue and yellow, compared with our three-colour trichromatic range of red, blue and green**. This means that they see reds and greens as shades of grey. But what they lose in colour complexity they gain in excellent long-range and low-light perception, wide field of vision and fast flicker-fusion movement detection.

Wolves (dogs' closest relatives) mainly hunt at dusk, so their vision is tuned to be most effective in low-light conditions, and dogs have inherited this trait. Their eyes have more rod receptors than cones, which means they can see light and shade better than colours. Unlike humans, they have an extraordinary **tapetum lucidum reflective layer at the back of their eyes that bounces light back to the retina, increasing the amount of light it can catch** and improving visibility. You can see this reflective layer when you take a flash photo of a dog: its shining eyes resemble the peepers of the devil himself. If you look carefully you may also spot that dogs have three eyelids: a top one, a bottom one, and a third one called a nictitating membrane. This is usually deployed while your dog is sleeping for extra eye protection, but you might see it as you wake up a sleepy hound.

Visual acuity is the ability to spot the gap between two lines from far away, and in this humans are easy winners: dogs need to be 6m (20ft) away from an object that we can see from 25m (82ft) away. Despite this, the rods that dominate their vision mean that their

perception of movement – even from a great distance – is much better than ours. This is handy for hunting small animals, and is helped by the positioning of their eyes either side of their muzzle, which gives most dogs a wider field of vision than humans. The only downside to this width is a relative lack of binocular vision (where two eyes visually overlap), which reduces their perception of depth.

Flicker-fusion rate refers to how much movement detail the eye and brain can process – for instance, high-definition TVs use 50–60 different images per second to create a smooth-looking moving picture. Movies used to be filmed at 24–25 frames per second, which gives the illusion of smooth movement as long as the camera doesn't pan too fast and the shutter speed isn't too short. **Dogs, however, process 70–80 frames per second, which means that in effect they see the world in more movement detail (and see TVs as a sequence of flickering images)** and have quick visual reactions. Compare that to a housefly's flicker-fusion rate of 400 frames per second and you'll understand why it's so hard to swat one – they effectively see the world in slow motion.

The Longest Ears

The longest dog ears belonged to Tigger, a Bloodhound from St Joseph in the USA. His right ear was 34.9cm (13.75in) and his left was 34.3cm (13.5in), according to *Guinness World Records*.

6.04 Dog taste, touch and hearing

Taste

Dogs' sense of taste is fully developed at birth, but nowhere near as sophisticated as ours. We have around 10,000 taste buds on our tongues, but dogs only have 1,700 (cats only have 470), and that's despite the fact their tongues are ridiculously long. In particular dogs have a low sensitivity to salt (possibly because a meat-rich diet naturally contains lots of salt), but a good sensitivity to sugars and acids and a strong dislike of bitterness. **Weirdly, they also have receptors on the tips of their tongues that are highly sensitive to the taste of water, and more so after eating salty or sugary foods.** It's thought this may help them in the wild when they need more water after eating substances that could dehydrate them.

Touch

Touch is the only other canine sense, along with taste, that's fully developed at birth. Dogs are highly sociable, so touch is an important tool for communication and interaction throughout their lives. They also enjoy being touched by humans, with many studies showing that gentle stroking reduces dogs' heart rate and blood pressure – and that humans get the same benefits, too. The most sensitive part of a dog's body is the muzzle, especially at the bases of the whiskers (known as vibrissae), which are packed with mechanoreceptors that respond to touch. The function of whiskers isn't fully understood, but as with cats, they are thought to help dogs understand their position in relation to an object that's too close to see.

Hearing

Although many dogs have large carpets of fur that flop over their ears, they have much better hearing than humans at high frequencies: dogs can hear sounds as high as 44,000Hz but we can only hear up to 19,000Hz. We outperform them at lower bass frequencies, though, detecting noises as low as 31Hz compared to their 64Hz. **Dogs' ears are also highly mobile, with 18 muscles controlling their position so they can quickly pinpoint the source of a sound**. Their ear structure helps them hear sounds from far away – around four times farther than humans.

Dalmatian Deafness

Thirty per cent of Dalmatians are deaf in one ear and 5% are deaf in both. It's caused by the extreme piebald gene, which is also responsible for their spotted coats and occasionally blue eyes. Dalmatians with larger spots are less likely to be deaf.

Chapter 07: Dog Talk

7.01 Why do dogs bark?

Barking isn't well understood. Dogs bark in different ways for lots of different reasons: boredom, fear, aggression, reaction to isolation, desire to play, need for human intervention, during play, as a protest, as a distress call, or simply when you say 'squirrel'. We don't even know whether barking is a form of communication or simply a reaction to a situation or experience. Why does this matter? Well, excessive barking is a common reason for people surrendering dogs to animal shelters. If we understood it more, perhaps we could help dogs lead better lives.

Wolves rarely bark, so barking may well have developed as part of domestication – perhaps barking dogs were selected over quieter ones because they could warn against intruders or predators. Or perhaps the close relationship between humans and dogs necessitated a means to alert us to their needs: 'I want food/drink/play/exercise/to pee'. This is a good example of conditioning – **if you feed your dog after he has barked, the association between bark and food will continue for both of you, so be careful**. There's even a decent argument for neoteny: if you choose for one trait (those cute, puppy-like floppy ears and big eyes), you tend to get a set of associated traits (puppies bark a lot) as a by-product.

Despite the lack of scientific consensus, you can still interpret your own dog's barks. Listen carefully, taking note of the tone, repetition, pitch and depth. You also need to note his – and your – body posture, the context and, crucially, the response (whether from another dog, a stranger, friend, or yourself). It takes a little time, but you should end up with a set of combinations that explain whether he's alarmed, hungry, angry, or whether there's a squirrel that needs annoying.

7.02 What does your dog hear when you talk?

Dogs can memorize scores of different words relating to objects and they also respond to an extensive set of orders such as 'sit', 'stay' and 'lie down'. They are excellent at understanding many of the human states of mind we communicate to them, and guide dogs even use basic semiotics (the use of signs to convey meaning). But none of this means that dogs understand what we think of as *language*: a complex, structured communication system complete with logical and lexical meaning, and grammatical rules. They struggle to identify specific words when we talk fast or in sentences, though they can often pick out words with strong sibilants such as 'squirrel' and 'sit', and long vowels such as 'heel' and 'walk'.

Some people think dogs respond to a tone of voice and don't register words at all. In 2017 a French bioacoustician (a scientist who studies sounds produced by and affecting living organisms) found that **women invariably use a slow, high-pitched, sing-song voice to address dogs, and that puppies played recordings of these voices responded strongly, barking and running toward the loudspeaker**. Some even performed the 'play bow' used to initiate play. On the other hand, most adult dogs who heard the same recording just looked at the speaker and ignored it. Researchers aren't entirely sure why – adult dogs still love to play, after all – though it's possible they had learnt that an invitation to play with no human present wasn't worth getting excited about.

Dogs are good at identifying our emotional states and intentions through body language and facial gestures, but they really excel when words are combined with intonation. A fascinating 2016 study

published in *Science* analyzed dogs' responses to certain phrases using dogs trained to sit in MRI scanners. It found that dogs' brains process language in the same way that we do: the right hemisphere processes emotion and the left hemisphere processes meaning. But the most fascinating discovery was that dogs only experienced happiness (or, scientifically speaking, neurological activity in their primary reward regions) if the words were combined with a tone of voice consistent with praise. When they heard 'good boy' in a flat, neutral voice, they recognized the phrase but not the praise, and their brains didn't register happiness. They were only happy when a praising intonation matched the words. This implies that **dogs identify words and intonation separately, but combine the two to interpret meaning, and are only genuinely happy when they *understand* the praise**.

Of course, whether or not dogs want to listen to you is another matter: a 2014 study published in *Behavioural Processes* revealed that dogs prefer petting to vocal praise by a wide margin.

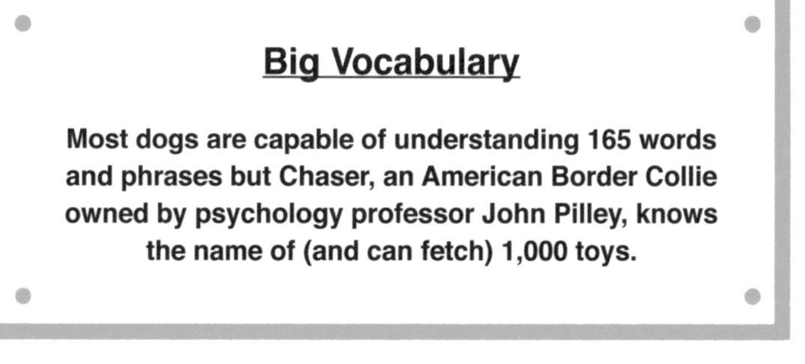

Big Vocabulary

Most dogs are capable of understanding 165 words and phrases but Chaser, an American Border Collie owned by psychology professor John Pilley, knows the name of (and can fetch) 1,000 toys.

7.03 What do growling, yodelling and howling mean?

Dogs make a surprisingly wide range of vocal sounds and, as with tail wagging (see p63), these are often context-specific. This is infuriating for researchers because a vocalization might mean one thing in one context, but something completely different in another – though owners and dogs usually train each other for mutual understanding.

Simple grunts are often heard in greetings and to express satisfaction (puppies regularly grunt when feeding and sleeping). Growls can signify aggression or defensiveness but are also common during play. A 2008 Hungarian study found that dogs understood specific growl meanings played to them through a speaker. When a dog eating a bone alone heard recordings of a group of dogs growling at each other while competing for their own bone, it tended to back away from its bone and stay away. This raises the possibility that dogs really can talk to each other (although possibly only about bones).

Some Dogs Can't Bark

Basenjis don't bark – they yodel and scream.

Whining and whimpering are common in puppies and adolescent dogs who are lonely, hungry, afraid or in pain, but can also be used by adult dogs to show submission, as a greeting or to seek attention. Whenever my mum came to visit, her gorgeous black Labrador, Daisy, would crawl around me, wagging her tail furiously and whimpering for several minutes. I took this to mean she loved me more than any human on the planet, but she might equally just have been over the moon about being let out of the car.

Yodelling and screaming is common in Basenjis, New Guinea singing dogs and dingoes, who can make these noises because their larynxes are narrower than most dogs', giving them more control over pitch. They may have been selected for this behaviour by humans who wanted them to sound like jackals or hyenas, and thereby ward off potential predators.

Howling is common in wolves but relatively rare in dogs other than wolf-like breeds such as Huskies and Malamutes. Wolves are thought to howl for several reasons: to locate members of their own pack, to claim their territory (which is often enormous), and to warn other packs to stay away. They also howl to assemble the pack for hunting or travel. These issues are often irrelevant to dogs, who instead may be triggered to howl by emergency sirens, airplanes or the music of Justin Bieber. Why, though, isn't understood. Dogs may also howl simply to get attention (and if you're trying to stop them from howling, attention, even negative attention such as scolding, is the last thing you should give them).

7.04 Can dogs talk to each other?

As well as barking, dogs have lots of other communication tools available to them: posture, facial gestures, ear position, fur erection, eye contact, peeing up lampposts. And dogs are much better at reading these signs than humans.

Olfactory (smell) communication is a dog's way of leaving – and learning – information about sex, health, age, social status and emotional state. **Your dog will spread his physiological message wherever he goes, using urine, poo, anal gland secretions and body odours to advertise his presence and potential as a mate**. He's spreading pheromones, chemicals that trigger a social response and behavioural change in other dogs.

But what about when two unfamiliar dogs meet? They use a complex interaction of body posture, eye contact and facial expression to communicate. The first thing they establish is whether or not one of them is dominant, and this starts with eye contact. The more dominant dog will establish eye contact first and keep it up for longer, while the submissive or younger dog will avert its gaze or avoid it altogether. Although this sounds a bit uncivilized to us, it's very useful: once dominance has been established, the dogs can socialize further, and the potential for aggression is minimized. On the other hand, if one dog doesn't assume submission, things can escalate pretty fast with bared teeth, growling, piloerection (bristling fur) and, if none of those work, physical attack.

Next, body posture comes into play, and again this relates to dominance and submission. The dominant dog will stand tall and straight, ears pointing up and forward, tail wagging at a high angle

and possibly also a small snarling twist to its face. The submissive dog will crouch, lower the angle of its tail, hold its ears back and sometimes show a 'submissive grin'. It may also try to lick the dominant dog and roll over on to its back to prove it's not a threat.

The 'play bow' is usually reserved for use between familiar dogs, and it's a clear invitation to play. Interestingly, humans often perform a clumsy version of this when they want to play with their dog. I know I do, bending down and patting my thighs while talking utter nonsense.

Tail wagging is important to dog interaction but is poorly understood. It's often a signal of friendship or positive excitement, but can also indicate that a dog is about to attack. Like vocalization, it seems to be a context-specific behaviour, meaning different things in different dogs in different situations (see p63). But, again, while researchers struggle to unearth a common language, dogs seem to communicate to each other pretty well.

Chapter 08: Dogs vs Humans

8.01 Cat people vs dog people

Or 'how to upset a large proportion of the world's population in 300 words'. Look, I know that there's a great variation in personality types, so I'm not saying that a dog owner like you is definitely an aggressive, overbearing, delusional egomaniac – I'm just saying that you *probably* are. Hang on – that's not coming out right. Look, I'm a dog lover, cat lover, gerbil lover and human lover so I'm not biased – it's just that a 2010 University of Texas study of self-identified dog lovers and cat lovers found that **cat folk tended to be less cooperative, conscientious, compassionate and outgoing than dog people and more likely to suffer from anxiety and depression**. But while cat people were more neurotic, they were also more open, artistic and intellectually curious than dog people. In 2015 researchers in Australia found that dog owners scored higher than cat owners on traits relating to competitiveness and social dominance, which matched their predictions (because dogs are more easily dominated, they assumed that their owners tended to be people who were more dominant). But they also found that cat owners scored just as highly as dog owners on narcissism and interpersonal dominance.

In 2016 Facebook published research on its own data (so, bear in mind it's specific to Facebook users, although the company does have a creepy ability to know a lot of things about people) and found that:

- Cat people are more likely to be single (30%) than dog people (24%).
- Dog people have more friends (well, they have more Facebook connections).
- Cat people are more likely to be invited to events.

Facebook also found that cat people are more literary in the books they mention (such as *Dracula*, *Watchmen*, *Alice in Wonderland*) and dog people more dog-obsessed and religious (*Marley and Me*, *Lessons from Rocky* – both about dogs – and *The Purpose Driven Life* and *The Shack* – both about God). Dog people like soppy movies about love and sex (*The Notebook*, *Dear John*, *Fifty Shades of Grey*), whereas cat people like death, hopelessness and drugs – with a bit of love and sex thrown in (*Terminator 2*, *Scott Pilgrim vs the World*, *Trainspotting*).

But Facebook's data gets really interesting (and scarily intrusive) when it comes to emotions. It truly seems to mirror the stereotypes of their animals, finding that **cat people are much more likely to express tiredness, amusement and annoyance in their online posts than dog people, whereas dog folk are more likely to express excitement, pride and 'blessedness'**.

Missing Organs

Dogs don't have an appendix. Neither do cats.

8.02 How much does a dog cost to run?

The cost of owning a dog is huge, and studies show that new owners woefully underestimate the likely financial burden. Annual costs are £445–£1,620 per year in the UK and $650–$2,115 in the USA – that's a total of £5,785–£21,060 or $8,450–$27,495 **over the course of an average dog's 13-year lifetime – PLUS the cost of buying your dog in the first place***. In comparison, Battersea Dogs and Cats Home reckons that a cat in the UK costs around £1,000 ($1,400) a year to look after.

Of course, the true cost of owning a dog depends on how much you *want* to spend. It's not unusual for a pedigree puppy to cost £3,000–£4,000 ($4,200–$5,600) just to buy, with higher pet insurance costs and grooming fees compared to a rescue animal from Battersea Dogs and Cats Home, which you can adopt for a donation of £155–£185 ($215–$260). Ongoing costs are what really hit the wallet, though. Food is likely to be your biggest expenditure, with amounts for British dog owners varying from £190 to £950 ($265 to $1,300) per year depending on the brands you buy and your dog's needs (specific nutritional requirements may mean expensive food). Another hugely important factor is pet sitting – who is going to look after Miss Scruffbag when you're at work or on holiday? In the UK professional pet sitters, dog walkers and holiday kennels can easily cost an extra £1,000 ($1,400) a year – though a neighbour or relative may always help you out of sheer love.

Other costs include regular vet's bills for vaccinations and check-ups, and kit such as bowls, collars, toys and a carrier for trips to the vet, not to mention set-up costs to cover microchipping, spaying/

neutering, buying a bed, kennel or crate. You also MUST insure your dog. I made the mistake of letting the insurance on my beloved elderly cat Tom lapse and his healthcare in the last year of his life cost me a crippling £3,000 ($4,200). Dog insurance in the UK easily costs £400 to £900 ($560–$1,250) a year depending on the dog, the policy and where you live (it's more expensive in big cities), but can shoot up for older dogs, and many companies simply won't insure elderly ones.

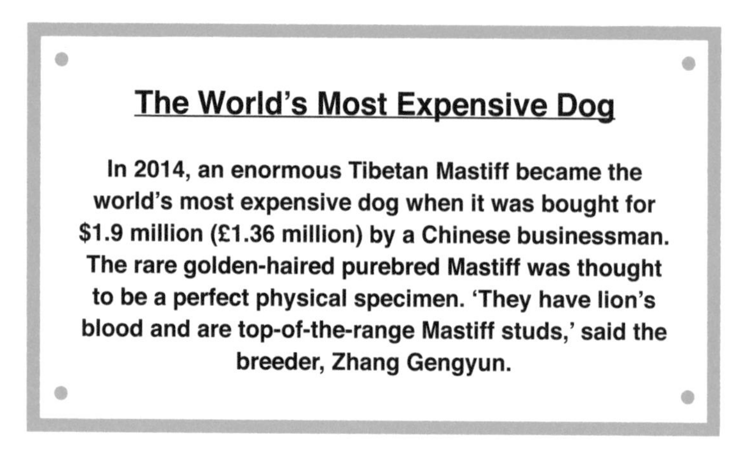

The World's Most Expensive Dog

In 2014, an enormous Tibetan Mastiff became the world's most expensive dog when it was bought for $1.9 million (£1.36 million) by a Chinese businessman. The rare golden-haired purebred Mastiff was thought to be a perfect physical specimen. 'They have lion's blood and are top-of-the-range Mastiff studs,' said the breeder, Zhang Gengyun.

*According to The Dog People, which estimates initial set-up kit and vet costs to be £730–£1,595 (but only $650–$2,115 in the USA).

8.03 Can you leave your fortune to a dog?

No. Yes. Sort of. You can't leave money or property to a dog because in law animals are property, and one piece of property can't own another piece of property. There are a couple of options, though. You can simply leave both your dog and some money to someone you trust, with an expectation they will use that money to look after the dog. However, they won't be legally obliged to do so, so it really does need to be someone you can count on. And don't think you can just name the dog in your will and leave everyone else to sort it out. You can say whatever you want in a will but that doesn't mean it's all enforceable.

If you're determined to look after your dog, and have the cash, you can set up a pet trust, which is a stronger but more expensive legal structure. You leave your dog, your money and, crucially, a legal obligation to a 'caretaker' to use that money to care for the dog, along with detailed instructions on how you want them to go about it. You also need to assign someone else to enforce those instructions and sue the caretaker if they fail to carry them out. It's a pretty big ask, so you can expect to pay both caretaker and trust enforcer a lot for the responsibility. Perhaps it's better just to leave your dog and money to an animal sanctuary or rescue organization instead?

When the notoriously mean American hotelier and real-estate magnate Leona Helmsley died in 2007, she tried to leave $12 million in her will to her dog, Trouble. But Trouble wasn't the only one due to benefit, because Helmsley also left instructions for the remainder of her entire trust, valued at $5–$8 billion, to go to help dogs. The only hitches were that her trustees had ultimate control over distributing the money, and that the request had not been incorporated into Helmsley's will or the trust documents. Funnily enough, the trust fund managed to find itself not legally bound to follow her wishes and in 2008 a judge ruled that Helmsley had been mentally unfit when she executed her will. As a result, most of the money left to Trouble went to her grandchildren, whom Helmsley had specifically disinherited, and none of the main programmes now operated by her trust have anything to do with dogs. Which leaves us with two conclusions: 1) Don't be mean. 2) Sort your bloody will out.

8.04 Do owners look like their dogs?

Several studies have found that **pets and their owners often resemble each other, and that it is uncannily easy to match complete strangers with their dogs – as long as the dogs are pure breed** (mixed breeds are harder to match).

There are several recurring themes: eye shape seems to be shared to some extent; women with long hair are more likely to choose dogs with long, floppy ears; and larger people tend to have fatter pets. Other research shows that the prevalence of pet obesity increases in line with that of human obesity.

Such findings seem to link with research indicating that humans prefer to pair up with people who look like them, and just as with dogs and their owners, strangers are remarkably good at identifying humans who are and aren't partners. It does feel slightly sad that we're *that* obvious.

Sheer Weirdness

Chinese Crested Dogs look very, very weird (don't get me wrong – this is a good thing, and if all of us were a little weirder, the world would be a better place). Powderpuff varieties have white human-like hair on their heads and tails but are otherwise dark-skinned and hairless. The effect is truly startling.

8.05 Are dogs good for your health?

Everyone *says* that owning a dog is good for your physical and mental health, so it must be true, right? On the surface, it certainly looks that way. A huge Swedish study published in 2017 tracked 3.4 million people aged between 40 and 80, and found that **dog ownership was associated with a 23% reduction in death from heart disease and a 20% lower risk of dying from any cause over a 12-year period**. In 2019 the American Heart Association found that 'Dog ownership was associated with a 33% lower risk of death for heart attack survivors living alone and 27% reduced risk of death for stroke survivors living alone, compared to people who did not own a dog. Dog ownership was associated with a 24% reduced risk of all-cause mortality and a 31% lower risk of death by heart attack or stroke compared to non-owners.'

Tapeworm Takeover

The Turkana people of north-western Kenya have the world's highest incidence of hydatid disease (caused by dog tapeworm) for a very specific reason. They live in unusually close proximity to their dogs, which play with and clean their children (including eating their faeces and vomit), lick plates and cookware clean, and poo inside compounds. Although hydatid disease is serious and can be fatal, water is scarce in the semi-arid region and the Turkana are reluctant to change their relationship with their dogs.

So, dogs are clearly good for your health, right? Oh God, you're going to hate me for this, aren't you? The answer is: not necessarily. The issue is in the words 'dog ownership was *associated* with'. A 2017 study by the RAND Corporation (a US non-profit research and development organization) found that pet ownership was, indeed, *associated* with health benefits, but that those health benefits may have other confounding variables (factors that warp results) – many of them tied to socioeconomic status. Better health rates seem more to do with the fact that pet owners tend to have bigger homes and greater household incomes, which both commonly relate to medical benefits. Then there's also the fact that dog owners are usually in better health to begin with: people with severe health problems are less likely to consider owning a pet that needs walking twice a day. Thus, the statistics are skewed before a study even begins. It's not so much 'I have a dog, therefore I'm healthy' as 'I'm healthy, therefore I'll get a dog'.

A study published in *Environmental Research* in 2019 even found that owning a pet was associated with a doubled risk of dying from lung cancer in women (although cats were much more to blame than dogs). And **according to the World Health Organization, every year 59,000 people die from rabies as a result of dog bites and millions are bitten by dogs**.

But what about claims of mental health benefits? Well, not many of those seem backed up by scientific research, either. There are quite a few shabbily designed studies that associate pet ownership with wellbeing but they are mostly self-reported and use tiny sample sizes – and in one case the research was 'carried out on Amazon Mechanical Turk', a crowdsourcing website, rather than by academic researchers. A more reliable-sounding 2020 study published in

the *International Journal of Environmental Research and Public Health* concluded that 'our findings are not in line with the notion that pet ownership generally has a health-benefiting effect'. Another in the *Journal of Veterinary Behavior* in 2014 said that 'dog owners perceived themselves as healthier – but not happier – than non-dog owners'. Yet another in *Anthrozoös* in 2019 found that **dog owners were less likely to report long-standing mental illness overall, but unmarried owners displayed increased odds of reporting long-standing mental illness**.

On the positive side, several studies have concluded that dog owners visit parks more often (and for longer) than non-dog owners, and the oxytocin release we get from interacting with dogs really ought to be good for our mental health. A 2012 Japanese study found that older dog owners exercised more hours per week than older non-dog owners. Another Swedish study published in 2015 reported that children aged between three and six years exposed to dogs in the first year of their life had a 13% lower incidence of asthma when they reached school age. A 2016 British study even said that reading aloud to dogs could lead to children improving their reading performance. It's not quite 'dogs are good for our health', but I'll take it.

Spaniel Rage Syndrome

Although it's a rare condition, some dogs are prone to idiopathic aggression: unpredictable outbursts of aggression and biting. It's thought to be a genetic trait that mainly affects Cocker and Springer Spaniels.

8.06 What does your dog do when you go out?

Most dogs experience some anxiety at being left alone in the house, and that shouldn't be surprising. We have bred dogs to be dependent on us, to want to be with us, to love us, and to rely on us for all the good things: food, drink, affection, companionship and play. And then we bugger off for the day.

Your dog's anxiety at being left on his own usually starts well before you go out of the door: dogs are highly skilled at reading human body language and will pick up on the particular tone of voice you use before leaving (even if you think you don't have one) and your body movements as you get your coat and check for keys, phone, wallet, bag. Your dog's stress levels peak soon after you've left. Often **the first 30 minutes are the worst as his heart rate rises, breathing speeds up, levels of the stress hormone cortisol increase, and – if your dog is prone to them – behaviours such as barking, whining and destruction kick in**. If the stress and boredom are particularly acute, he may experience saliva production, urination, repetitive pacing and even self-mutilation.

You can train a dog to cope with being left alone, but it's best to tackle it from an early age. Start with short periods of isolation so your dog knows that being left alone doesn't mean he is being abandoned, and that you will return. As you gradually increase the amount of time you're away, your dog should become desensitized to the experience. One tip is not to make a fuss of him before you leave, as this will only increase his anxiety. Another tip is to have a box of distracting treats that only appears when you leave the house

and is packed away as soon as you return. And whatever you do, don't punish him if he destroys something or goes to the toilet in the house while you are away. **Punishment is rarely, if ever, constructive in training**; your dog doesn't make the connection between the act and the punishment, and piling one form of distress on top of another is only likely to make him more anxious about you going out.

You may think your dog is resilient enough to cope with being left alone, but that's not necessarily the case. Just search 'dog left alone' on YouTube to see how distressed a tough-looking Rottweiler can get (and how he drinks from the toilet bowl) when his owner's not around. Warning: it's heartbreaking stuff.

Labradoodles (Labrador Retriever/Standard Poodle crossbreeds) are very popular in Australia where they are on their way to becoming a pedigree breed. The hybrid was founded in 1989 by Wally Conron of Guide Dogs Victoria in response to the need for a guide dog of a vision-impaired woman whose husband had dog allergies. British speed record breaker Donald Campbell had previously bred one in 1949, and named it a Labradoodle.

8.07 The climate impact of your dog

Owning a dog has many benefits, but it also places a huge burden on the environment. Dog faeces often ends up in landfill, their presence disrupts wildlife habitats, and they reduce biodiversity by either attacking or scaring other animals away. But more important by far is the impact of all the food they eat, which requires energy to produce, harvest, package and transport, and affects our ability to feed ourselves.

A 2017 UCLA study concluded that **US dogs and cats consume about 19% of the amount of dietary energy that humans do and thereby add an extra 19% to the ecological burden that we already place on our planet**. They also consume about 33% as much animal-derived energy, produce about 30% as much faecal matter, and the land, water, fossil fuel, phosphate and biocides used to sustain them are responsible for about 25–30% of all environmental impacts from animal production. The author acknowledges that pet food is invariably made from meat by-products consumed by humans, but counters that if dogs can eat those, humans should be able to, too. Admittedly, tripe, lungs and entrails aren't enjoyed on a large scale by humans, so doing so would require a big cultural shift; that said, they can be delicious (I'm particularly partial to a bit of lung).

The study acknowledges that: 'People love their pets. They provide a host of real and perceived benefits to people …'. Nevertheless, we should be aware that our pets represent a significant ecological burden, which we need to take into account when we try to mitigate our own impact. This opens up a world of moral and ecological relativity where we have to balance unquantifiable

emotional impact (I love my dog *so* much) with quantifiable climate impacts (my dog eats another 19% of my dietary energy requirements), and this can lead us into difficult territory. After all, one of the most effective ways of reducing your CO_2-equivalent emissions is to cut down on the number of children you have: having one fewer child saves 58.6 tonnes (64.6 tons) of CO_2-equivalent emissions per year (changing to a plant-based diet only saves 0.8 tonnes (0.9 tons) of CO_2-equivalent emissions per year). Of course, we love our kids, and it's both impossible and horrific to quantify whether or not all that extra love in the room outweighs the downsides. There is definitely a balance to be achieved, and discussions to have, but is it just a short jump from cutting down on family pets to a one-child policy?

Legendary Dogs

Psycho Ren

Chihuahua is a huge and mountainous area in north-western Mexico, the largest state in the country and 2% larger than the entire UK. So it must be galling that the first 99 results for 'Chihuahua' on Google are about the smallest dog breed in the world. The Long-haired and Short-haired varieties are seen as two separate breeds and the most famous examples are probably the brilliantly psychotic Ren Höek (from the animated series *The Ren & Stimpy Show*) and Gidget, mascot for the Taco Bell fast-food chain from 1997 to 2000 and Bruiser's mum in *Legally Blonde 2*. Both were Short-haired Chihuahuas.

Chapter 09: Dogs vs Cats

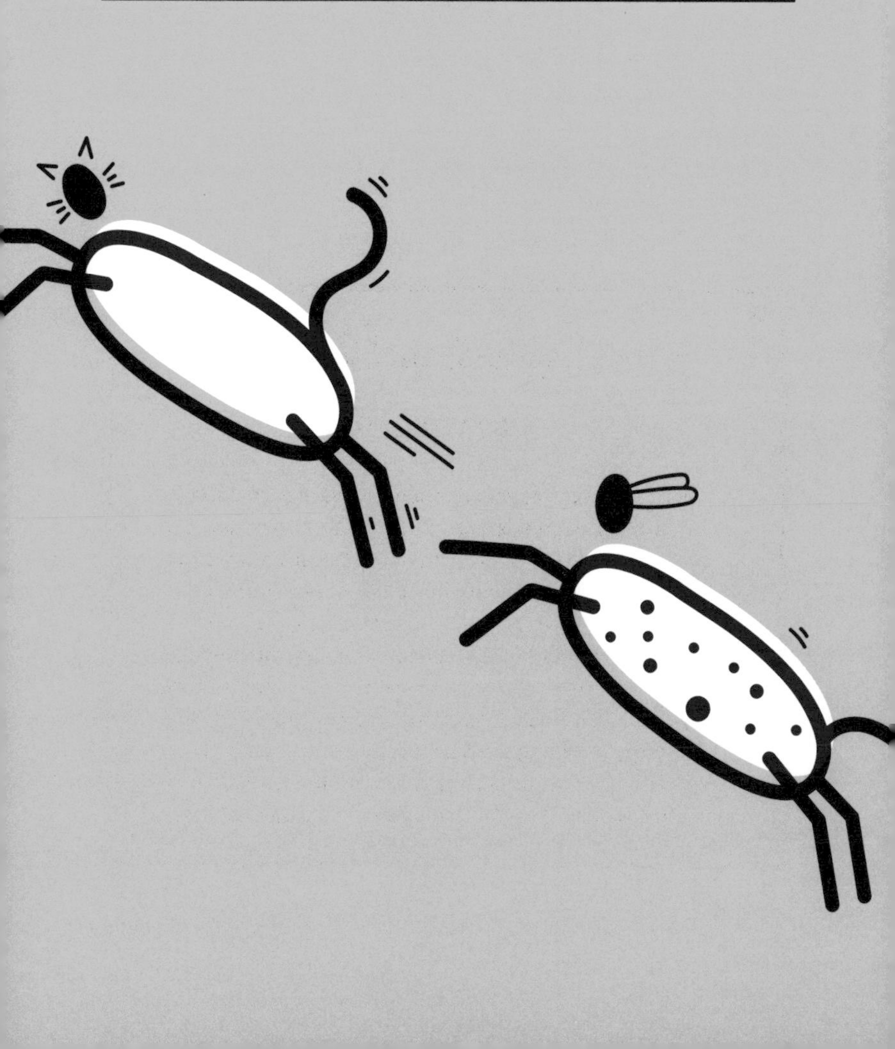

9.01 Can one species be better than another?

Before diving headlong into the great Dog vs Cats debate, let's pause for a conceptual biology moment. Don't worry – this shouldn't hurt too much.

With our opposable thumbs, powers of abstract thought and fabulous music taste, we humans like to think we're superior to all the other species on Earth. Apes and dolphins may not be far behind, but earthworms and plankton? Pah! Look at what we've achieved: our impact on the Earth is so great that the Holocene era (the 12,000 years since the last ice age during which human civilization developed) is now thought to be over, replaced by the Anthropocene, an era defined by humans' pre-eminent impact on the planet. Following the invention of the spork, selfie stick and Justin Bieber, it's fair to say that species don't come much more perfectly formed than us. Yeah! Go, humans! Except, of course, the Anthropocene is defined by disastrous markers starting with radioactive pollution in the 1950s, a striking acceleration of CO_2 emissions, mass deforestation, biodegradation, war, inequality and the global mass extinction of species.

On the other hand, earthworms' ancestors survived five extinction events and existed for 600 million years compared to humans' paltry 200,000 years. Darwin thought that earthworms played one of the biggest starring roles in the history of the world, ploughing and fertilizing our soil and making it possible for us to grow food. And plankton? Well, just look at the numbers: 7.8 billion humans is pretty paltry compared to the SAR11 plankton population at 2.4 x 10^{28}. That's 24,000,000,000,000,000,000,000,000,000 plankton to you, squire.

So, asking whether dogs are better than cats is generally considered a fool's game, a bit like asking 'What's better: a tree or a whale?' A tree excels at being a tree and a whale excels at being a whale. An earthworm isn't better or worse than a human – it's excellent at being a terrestrial hermaphroditic invertebrate that respires through its skin and lives underground. And even then, species are thought never to be at an evolutionary optimum, but always in some form of adaptation in relation to their situation. Dogs' and cats' domestication is particularly interesting: in evolutionary terms they are wild hunting predators that have only relatively recently moved into our houses, so are probably only at the start of an adaptation phase. Check in again in half a million years and they may be biologically very different. And looking at the way the Anthropocene is unfolding, their beloved humans may no longer exist at all.

Dog-lead Injuries

Dog-lead injuries in the USA run at a level of 63.4 per 1 million people. The most common are of the pull-followed-by-trip/tangle variety. And one-third of them happen in the home. Which is weird if you think about it.

9.02 Dogs vs Cats: social and medical

The previous pages went to great lengths to explain why comparing dogs with cats goes against conceptual biological principles. But where's the fun in that? Come on, let's play Dogs vs Cats!

Popularity

Dogs are way more popular than cats in the UK* (although statistics do vary wildly).

23% of households own at least one dog

16% of households own at least one cat

Winner: DOGS

Love

Owners of both species are pretty wild about their pets, but which animal loves us more? Neuroscientist Dr Paul Zak analyzed saliva samples from dogs and cats to find out which contained more oxytocin (the hormone associated with love and attachment) after playing with their owners. Cats' oxytocin levels increased by an average of 12% but dogs' rose by a whopping 57.2%. That's a six-times greater increase. Dr Zak even twisted the knife, noting: 'It was a nice surprise to discover that cats produce any [oxytocin] at all.'

Winner: DOGS

Intelligence

Dogs' brains, at an average 62g (2oz), are bigger than those of cats', which average 25g (0.9oz). But that doesn't necessarily make them cleverer – sperm whales have brains six times the size of humans',

but are still considered less intelligent because, among mammals, we have the largest cerebral cortex (the area responsible for higher-order functions such as information processing, perception, sensation, communication, thought, language and memory) relative to the size of our brains. Another measure of intelligence is the number of neurons in an animal's cerebral cortex. Neurons are fascinating because they have a high metabolic cost (they use a lot of energy to keep running) so the more neurons we have, the more food we need to consume, and the more metabolic machinery we have to run to turn that into usable fuel. Because of this, each species only has as many neurons as they absolutely need, and a paper published in *Frontiers in Neuroanatomy* found that **dogs have more neurons in their cerebral cortex than cats – about 528 million compared with 250 million.** Humans trump them both with 16 billion, though. The researcher who developed the measuring method said 'I believe the absolute number of neurons an animal has, especially in the cerebral cortex, determines the richness of their internal mental state … dogs have the biological capability of doing much more complex and flexible things with their lives than cats can.'

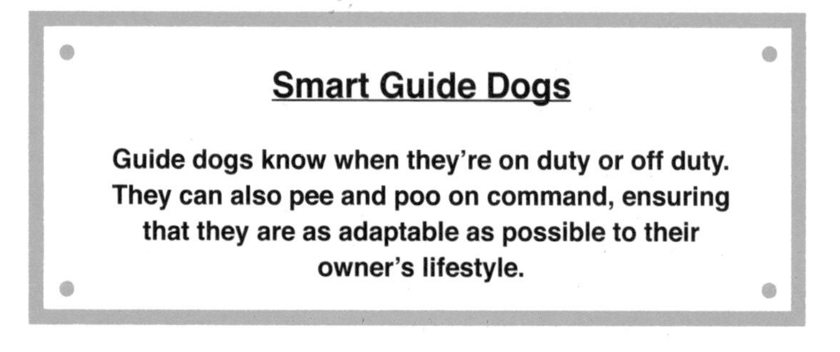

Smart Guide Dogs

Guide dogs know when they're on duty or off duty. They can also pee and poo on command, ensuring that they are as adaptable as possible to their owner's lifestyle.

*Pet Food Manufacturers' Association Pet Population 2020 report

The brain issue really depends on what's most important to a particular animal – dogs are pack animals, so need more communication skills, and these are centred in the frontal lobe and temporal lobe, whereas cats are solitary hunters and may need more motor function skills, centred in the frontal lobe's motor cortex, to control escape abilities. Of course, it's one thing having lots of neurons, but a better measure of intelligence may be what you do with them. Japanese scientists tested dogs and cats' memories and found no significant difference, but in terms of problem-solving capacity, dogs tend to rely on their owners to sort stuff out, while cats try to solve problems on their own.

Winner: DOGS

Legendary Dogs

Rin Tin Tin

This German Shepherd became a huge Hollywood movie star after he was rescued from a French World War I battlefield by Corporal Lee Duncan, an American army aerial gunner, in September 1918. He was named after one of a pair of good luck charms that French children would give to US soldiers (the other was called Nénette). After much perseverance, Duncan landed Rin Tin Tin a part in a film and he went on to appear in 27 movies, most of them silent. He was so popular in his first 1923 starring role that he was credited with saving Warner Bros from bankruptcy and is thought to have won the 1929 Oscar for Best Actor before the Academy, keen to keep its credibility, re-ran the vote to ensure that a human won.

Convenience

Cats are cheaper to buy, keep, feed and care for. They are independent, don't need walking, and can be left alone for much longer than dogs. They will happily poo and pee outside, and usually not in your own garden (good for you, not so good for your neighbours). And dogs? Dogs are not convenient.

Winner: CATS

Sociability

Cats are solitary and territorial animals but get physiological benefits from contact with humans. Dogs are sociable with each other, although prefer the company of humans. They respond to many human commands and requests, and enjoy physical contact – which, like with cats, provides physiological benefits.

Winner: DOGS

Eco-friendliness

Cats kill millions of birds every year (although the impact and exact number are strongly disputed) and both dogs and cats may reduce biodiversity. Dogs, on the other hand, have a larger ecological footprint: feeding a medium-sized dog requires 0.84 hectares (2.08 acres) of land a year compared to 0.15 hectares (0.37 acres) for a cat.

Winner: CATS (just)

Health benefits

Both dog and cat owners get positive hormonal benefits from contact with their pets (which helps lower stress levels) and have better immunoglobulin levels than non-owners, potentially offering higher protection from gastrointestinal, respiratory and urinary tract

infections. However, many of the grander health claims associated with pet ownership have been questioned in more recent studies. Dog owners tend to exercise more than cat owners and non-pet-owners, which is likely to lower cardiovascular risks and increase survival rates after heart attacks. But these benefits are countered by the facts that every year in the UK 250,000 people have to visit minor injury and emergency units after suffering dog bites and two to three people die as a result of dog attacks. Globally, rabid dogs cause the death of around 59,000 people a year, according to the World Health Organization.

Winner: CATS

Trainability

Dogs: the average dog can be trained to memorize 165 words and actions, catch a ball, sit, give a paw, jump, heel, lie on their bed, roll over, wait, and – at a push – stop humping Auntie Gladys's leg.

Cats: HAHAHAHAHAHAHA.

Winner: DOGS

Usefulness

Mouse-hunting is extremely useful – to the handful of people who own grain silos/farms or have a mouse infestation. To the rest of us, it's a bit irritating. Bird-hunting, on the other hand, is downright out of order. And that's about all cats offer – other than the obvious joy they give us whenever they can be bothered to seek our attention. Dogs, on the other hand, are useful for hunting, sniffing out contraband and explosives, tracking in the wilderness, diagnosing diseases, rescuing lost or trapped people, guiding the visually impaired, herding sheep, guarding houses, chasing criminals. I'll stop now – you know what I'm saying.

Winner: DOGS

B
BOMB SQUAD

9.03 Dogs vs Cats: physical head-to-head

Speed

Cheetahs are the fastest animals on land, able to run at 117.5km/h (73mph). But your cat, bless him, isn't a cheetah. If he can be bothered, he's likely to be able to achieve 32–48km/h (20–30mph) for short bursts. This compares shabbily with greyhounds' maximum speed of 72km/h (45mph), but pretty well against a lumbering Golden Retriever at 30km/h (19mph).

Winner: DOGS

Stamina

Dogs are hands-down winners here. Cats are ambush predators, capable of patiently stalking their prey for hours before sprint-pouncing. Dogs aren't built for sprinting but for long-range aerobic endurance chasing (much like myself, as it happens). Humans have co-opted this ability for travelling across ice and snow, and the stamina that sled dogs show is extraordinary – animals in the Iditarod Trail Sled Dog Race cover 1,510km (940 miles) of sparsely populated Alaska in eight to 15 days.

Winner: DOGS

Hunting ability

Despite being fed regular meals, almost all domestic cats retain both the urge and the skills to hunt, often bringing home mice and birds in differing states of completeness. Conversely, most dogs have a chase instinct, but the vast majority have a hunting ability best described as laughable – unless bred specifically for the task. My dog will chase my cat across the garden at top speed, but once he corners her, the

fun's over, and he wants her to run again. For her part, she just wishes he was dead.

Winner: CATS

Number of toes

What do you mean I'm scraping the barrel? Toes are important. Polydactylism is relatively common in cats, but rare in dogs.

Winner: CATS

Evolution

A 2015 study published in *Proceedings of the National Academy of Sciences* revealed that members of the cat family have in the past been better at surviving than members of the dog family. Dogs originated in North America about 40 million years ago and by 20 million years ago the continent was home to over 30 species. It could have been so many more, but for cats. The researchers found that **cats played a significant role in making 40 dog species extinct by outcompeting them for the available food**, whereas there was no evidence that dogs wiped out a single cat species. Different hunting methods may have been to blame for dogs' failure, as well as cats' claws, which are retractable and so remain constantly sharp. By contrast, dogs' claws don't retract and are usually blunter. Whatever the cause, the report stated that 'Felids must have been more efficient predators', which means at some level they were clearly *better.*

Winner: CATS

Chapter 10: Dog Food & Drink

10.01 Can dogs be vegetarians?

Thousands of years of cohabitation with humans and feeding on our leftovers has changed dogs from hypercarnivores (who rely on meat) into omnivores (who can eat anything), albeit with a carnivorous bias. Like wolves, dogs have short gastrointestinal tracts best suited to meat consumption, but unlike wolves, dogs can break down carbohydrates. Most dogs will, frankly, eat anything that fits in their mouths, from vegetables and cheese to shoes and toys. The difference is that dogs, like humans, produce the enzyme amylase, which breaks down plant starches and allows them to extract nutrition from grains. This ability probably developed because domesticated dogs constantly ate scraps and leftovers, which changed their digestive system.

With their sharp, pointy teeth and relatively short digestive tract, dogs are physically more suited to a carnivorous diet (by contrast, humans have an enormous digestive tract, specifically for digesting complex carbs and fibrous vegetables). Their ideal dietary breakdown is 56% protein, 30% fat and 14% carbohydrate. They also require the amino acids taurine and arginine as well as vitamin D, which they would normally get from animal flesh, but can be added to their food as supplements.

It is possible to feed dogs a vegetarian diet using commercial high-protein vegetarian and vegan pet foods supplemented with amino acids and vitamins, but you need to take great care to meet their nutritional needs. According to the president of the British Veterinary Association: 'It is theoretically possible to feed a dog a vegetarian diet, but it's much easier to get it wrong than to get it right.'

10.02 What's all the fuss about bones?

Why do dogs love eating bones? It looks like a lot of effort for small returns. Well, bones are packed with nutrient-dense and calorie-dense bone marrow, a soft, fatty, spongy tissue that's the main production site for new blood cells in mammals and birds. Dogs get lots of great nutrition from this marrow, but many also eat the bone itself. This seems odd as it's tough and difficult to break down, but plenty of dogs thoroughly enjoy the process, often spending hours gnawing away at a large bone until there's nothing left.

The reason dogs love bones *so* much could be down to their wolf ancestors. In late winter when food is scarce, the large mammals on which wolves prey are usually carrying less fat. The wolves most likely to survive until the next season are those able to get maximum nutrition out of their kill, and the last reservoir of dense calories is the fat in bone marrow. That means any wolf with a quirky love of chewing on bones has a better chance of surviving than one that doesn't. **Bones are also an excellent calorie-storage system: bone marrow can survive well inside an intact bone, making it particularly suited to being buried and dug up again later when a dog is hungry (and the rest of its pack has gone elsewhere)**.

A word of advice: only give your dog raw bones, never cooked ones. Cooking dries the hard compact bone outer shell, making it brittle and more likely to create sharp fragments that can harm dogs' mouths and intestines.

10.03 Why are dogs so greedy?

Canine obesity in developed countries ranges from 34% to 59%, and can lead to early death and multiple health problems. But as with humans, it's not simply greediness and lack of self-control – it's in their genes.

Dogs have retained the ability of their wolflike ancestors to scoff vast amounts of food in one sitting. Wolves are highly active pack animals that work together to hunt large mammals, but as soon as they bring down their prey, each individual has to compete with the rest of the pack to eat as quickly as possible and get a decent share of the kill. These kills may not happen often, especially in winter, so it's important wolves can eat enough food in one sitting to survive until the next one, which may be days or even weeks away. Domestic dogs have kept this scoffing instinct, despite the fact they are fed every day and many are relatively inactive, leading to a huge increase in canine obesity.

It's thought that **almost two-thirds of dogs in developed countries are overweight, and the worst-affected are Labradors**. Why Labradors? A 2016 study published in *Cell Metabolism* found the likely answer: a quarter of Labradors carry a copy of a gene mutation called POMC that encodes proteins that help switch off the hunger impulse after eating. This mutation is significantly more common in dogs selected to become assistance dogs such as guide dogs rather than pets, almost certainly because the most trainable Labradors are the ones that respond most to food-based training. Because humans select and breed the dogs that are most trainable (as they're so food-motivated) and so more useful, we have inadvertently made the problem worse.

10.04 What's in dog food?

In 2020 the global pet food market was worth £54.8 billion ($74.6 billion), while the UK market alone was worth £2.9 billion ($3.7 billion). The first commercial pet food was launched in England in the 1860s by James Spratt, an American entrepreneur who had travelled to London to sell lightning conductors but got distracted, so the story goes, when he was given some inedible ship's biscuits for his dog. He spotted a gap in the market and came up with his Meat Fibrine Dog Cakes. Yummy. He was hugely successful, first in the UK and later in the USA, and one of his early UK employees was Charles Cruft, who eventually left to set up Crufts dog show.

Despite the horror stories, pet food companies don't just throw anything into dog food. It's a highly regulated industry and some standards are surprisingly high: the animals used as ingredients have to pass vet inspections to ensure they are fit for human consumption at time of slaughter. Pets, road kill, wild animals, lab animals and fur-bearing animals aren't allowed, nor is meat from sick or diseased animals. Dog food is usually a mixture of cow, chicken, lamb and fish off-cuts, derivatives and by-products from food made for human consumption. This usually includes **liver, kidneys, udders, tripe, trotters and lungs, which might not sound particularly appetizing to you, but dogs love them** (when wolves kill prey in the wild they often eat the lungs, stomach lining, liver, heart and kidneys before the big muscle groupings). Importantly, it also means that nothing usable from a slaughtered animal goes to waste.

Although commercial dog food is mostly meat, grains such as corn and wheat are also often added, along with nutritional additives such as extra taurine (an amino acid that dogs can't make themselves),

vitamins A, D, E, K and various minerals. There's a trend towards 'grain-free' food, but do check with your vet – dogs need to get fibre from somewhere.

Wet dog food is generally made from meat and meat derivatives mixed with cereals, vegetables and nutritional additives such as taurine, and then cooked into a meatloaf before being cut into chunks and mixed with either jelly or gravy. It's then put into **cans, trays or pouches that are re-cooked in a retort (a huge pressure cooker) at 116–130°C (241–266°F) to kill bacteria**, leaving the sealed pack surprisingly sterile and with a long shelf life. The cans are then cooled before labelling.

Dry dog food (or kibble) is more interesting. As with wet food, it starts with a mixture of meat and meat derivatives but these are usually cooked and ground into dry powders before being mixed with cereals, vegetables and nutritional additives. Water and steam are added to make a hot, thick dough and this is pushed through an extruder – a huge screw thread that compresses and heats the dough – before it's forced through a small nozzle called a die (cheese puffs are made in much the same way) and chopped into shapes by revolving blades as it squirts out. This heating degrades some of the nutrients in the meat so more of these need to be added again later. The change of pressure as the cooked dough comes out makes it puff up into kibbles and these are heated to dry out before being sprayed with flavourings and nutritional additives to replace those degraded by the whole process.

Recently there has been a strong trend towards raw meat diets. If that's your bag, do be careful how you go about it and, as ever, check with a vet rather than read opinion pieces that aren't based on research.

10.05 What foods are poisonous to dogs?

Dogs should NEVER be fed:

1. **Chocolate** – the stimulants theobromine and caffeine are dangerous to them.

2. **Onions, chives and garlic** – can cause gastrointestinal irritation and red blood cell damage.

3. **Coffee** – again, theobromine and caffeine are dangerous.

4. **Xylitol** – found in many chewing gums, these can cause hypoglycaemia and liver failure.

5. **Avocados** – a compound in them called persin can cause vomiting and diarrhoea.

6. **Grapes and raisins** – can cause severe liver damage.

7. **Macadamia nuts** – they contain toxins that affect dogs' muscles and nervous systems.

8. **Corn on the cob** – cobs can get stuck in the digestive tract.

The theobromine and caffeine in chocolate and coffee can affect a dog's nervous system, increase heart rate, cause kidney failure and decrease body temperature. Dogs will react differently to chocolate depending on their size, sensitivity to stimulants, and the proportion of theobromine and caffeine in the chocolate they've eaten (darker chocolate contains more than milk chocolate). An early sign of poisoning is excessive drooling, vomiting and diarrhoea, and if you suspect your dog has eaten chocolate, it's best to contact your vet as soon as possible.

Puppies Can be Born Bright Green

In 2020 one of a litter of Italian puppies was born with a distinct green tint to its fur, while a German Shepherd in the USA gave birth to a litter of eight pups, one of which was lime-green. The owners named it Hulk. The strange phenomenon may be the result of the puppies coming into contact with either meconium (a green tar-like substance that lines newborn babies' intestines and usually makes up their first poo) or a green-pigmented compound called biliverdin from the mother's placenta. The colour usually fades after a few weeks.

Sources

I read an ocean of books, articles and research papers whilst writing *Dogology*, and I am indebted to the wonderful authors of all of them (apologies that only a tiny fraction are listed here) despite a wide range of results, some of them downright contradictory. But that's the nature of scientific research – as the methodology changes, so does the nature of the results, and science communicators like me have to read as widely as possible, assess the relevance and context, and tread a path through the information, hoping that we aren't straying from the truth. I have tried my damndest to be very clear whether I'm reporting either scientific research or opinion, even if that opinion comes from veterinary professionals. There's so much more to learn about dogs, and every new piece of research helps us understand them more and care for them better.

General

'Rabies: Epidemiology and burden of disease'
who.int/rabies/epidemiology/en/

'Meta analytical study to investigate the risk factors for aggressive dog-human interactions' (DEFRA)
sciencesearch.defra.gov.uk/Default.aspx?Menu=Menu& Module=More&Location =None&Completed=0&ProjectID=16649

'Pet Population 2020' (PFMA)
pfma.org.uk/pet-population-2021

'PDSA Animal Wellbeing (PAW) Report 2020' (PDSA/YouGov)
pdsa.org.uk/media/10540/pdsa-paw-report-2020.pdf
The 2020 PDSA (People's Dispensary for Sick Animals) survey via YouGov is startlingly comprehensive and has very different results to the PDSA with a larger sample size – showing 10.9 million cats to 10.1 million dogs. But the way it presents the data made me a little wary.

'Pet Industry Market Size & Ownership Statistics' (American Pet Products Association)
americanpetproducts.org/press_industrytrends.asp

'Pet ownership Global GfK survey' (GfK, 2016)
cdn2.hubspot.net/hubfs/2405078/cms-pdfs/fileadmin/user_upload/country_one_pager/nl/documents/global-gfk-survey_pet-ownership_2016.pdf

'Ancient European dog genomes reveal continuity since the Early Neolithic' by Laura R Botigué *et al*, *Nature Communications* 8, 16082 (2017)
nature.com/articles/ncomms16082

'Dogs Trust: Facts and figures'
dogstrustdogschool.org.uk/facts-and-figures/

'In what sense are dogs special? Canine cognition in comparative context' by Stephen EG Lea & Britta Osthaus, *Learning & Behavior* 46 (2018), pp335–363
link.springer.com/article/10.3758%2Fs13420-018-0349-7

2.01 A brief history of the dog

'Dog domestication and the dual dispersal of people and dogs into the Americas' by Angela R Perri *et al*, *Proceedings of the National Academy of Sciences of the United States of America* 118(6) (2021), e2010083118
pnas.org/content/118/6/e2010083118

2.02 Is your dog basically a cute wolf?

'Dietary nutrient profiles of wild wolves: insights for optimal dog nutrition?' by Guido Bosch, Esther A Hagen-Plantinga, Wouter H Hendriks, *British Journal of Nutrition* 113(S1) (2015), ppS40–S54
pubmed.ncbi.nlm.nih.gov/25415597/

'Social Cognitive Evolution in Captive Foxes Is a Correlated By-Product of Experimental Domestication' by Brian Hare *et al*, *Current Biology* 15(3) (2005), pp226–230
sciencedirect.com/science/article/pii/S0960982205000928

2.03 How were dogs domesticated?

'Dog domestication and the dual dispersal of people and dogs into the Americas' by Angela R Perri *et al*, *Proceedings of the National Academy of Sciences of the United States of America* 118(6) (2021), e2010083118
pnas.org/content/118/6/e2010083118

'A new look at an old dog: Bonn-Oberkassel reconsidered' by Luc Janssens *et al*, *Journal of Archaeological Science* 92 (2018), pp126–138
sciencedirect.com/science/article/abs/pii/S0305440318300049

'Dogs were domesticated not once, but twice… in different parts of the world'
ox.ac.uk/news/2016-06-02-dogs-were-domesticated-not-once-twice%E2%80%A6-different-parts-world#

2.04 Why do humans love dogs?

'Oxytocin-gaze positive loop and the coevolution of human-dog bonds' by Miho Nagasawa *et al*, *Science* 348: 6232 (2015), pp333–336
science.sciencemag.org/content/348/6232/333

'Neurophysiological correlates of affiliative behaviour between humans and dogs' by JSJ Odendaal & RA Meintjes, *The Veterinary Journal* 165:3 (2003), pp296–301
sciencedirect.com/science/article/abs/pii/S109002330200237X?via%3Dihub

'Oxytocin enhances the appropriate use of human social cues by the domestic dog (*Canis familiaris*) in an object choice task' by JL Oliva, JL Rault, B Appleton & A Lill, *Animal Cognition* 18 (2015), pp 767–775
link.springer.com/article/10.1007/s10071-015-0843-7

'How dogs stole our hearts'
sciencemag.org/news/2015/04/how-dogs-stole-our-hearts

2.05 Why do dogs love humans?

'Structural variants in genes associated with human Williams-Beuren syndrome underlie stereotypical hypersociability in domestic dogs' by Bridgett M vonHoldt *et al*, *Science Advances* 3:7 (2017), e1700398
advances.sciencemag.org/content/3/7/e1700398

'Neurophysiological correlates of affiliative behaviour between humans and dogs' by JSJ Odendaal & RA Meintjes, *The Veterinary Journal* 165:3 (2003), pp296–301
sciencedirect.com/science/article/abs/pii/S109002330200237X?via%3Dihub

'For the love of dog: How our canine companions evolved for affection'
newscientist.com/article/mg24532630-700-for-the-love-of-dog-how-our-canine-companions-evolved-for-affection/

3.03 Why do dogs poo facing north–south?

'Cryptochrome 1 in retinal cone photoreceptors suggests a novel functional role in mammals' by Christine Nießner *et al*, *Scientific Reports* 6, 21848 (2016)
nature.com/articles/srep21848

'Dogs are sensitive to small variations of the Earth's magnetic field' by Vlastimil Hart *et al*, *Frontiers in Zoology* 10:80 (2013)
frontiersinzoology.biomedcentral.com/articles/10.1186/1742-9994-10-80

'Pointer dogs: Pups poop along north-south magnetic lines'
livescience.com/42317-dogs-poop-along-north-south-magnetic-lines.html

'Magnetoreception molecule found in the eyes of dogs and primates'
brain.mpg.de//news-events/news/news/archive/2016/february/article/magnetoreception-molecule-found-in-the-eyes-of-dogs-and-primates.html

3.04 How many hairs are there on your dog?

'Weight to body surface area conversion for dogs'
msdvetmanual.com/special-subjects/reference-guides/weight-to-body-surface-area-conversion-for-dogs

3.07 Why do dogs make such a mess when drinking?

'Dogs lap using acceleration-driven open pumping' by Sean Gart, John J Socha, Pavlos P Vlachos & Sunghwan Jung, *Proceedings of the National Academy of Sciences of the United States of America*, 112(52) (2015), 15798–15802
pnas.org/content/112/52/15798

4.01 Why do dogs fart (but cats don't)?

'The difference between dog and cat nutrition'
en.engormix.com/pets/articles/the-difference-between-dog-t33740.htm

'Digestive Tract Comparison'
cpp.edu/honorscollege/documents/convocation/AG/AVS_Jolitz.pdf

4.02 The science of dog poo

'Dog Fouling'
hansard.parliament.uk/Commons/2017-03-14/debates/EB380013-5820-42A0-A7B9-29FF672000CE/DogFouling

4.04 Why do male dogs raise their legs to pee?

'Urine marking in male domestic dogs: honest or dishonest?' by B McGuire, B Olsen, KE Bemis, D Orantes, *Journal of Zoology* 306:3 (2018), pp163–170
zslpublications.onlinelibrary.wiley.com/doi/abs/10.1111/jzo.12603?af=R

4.09 Is it bad to let a dog lick your face?

'The Canine Oral Microbiome' by Floyd E Dewhirst *et al*, *PLOS ONE* 7(4) (2012), e36067
journals.plos.org/plosone/article?id=10.1371/journal.pone.0036067

4.10 Why do dogs sniff each other's bums?

'When the nose doesn't know: canine olfactory function associated with health, management, and potential links to microbiota' by Eileen K Jenkins, Mallory T DeChant & Erin B Perry, *Frontiers in Veterinary Science* 5:56 (2018)
ncbi.nlm.nih.gov/pmc/articles/PMC5884888/

'Dyadic interactions between domestic dogs' by John WS Bradshaw & Amanda M Lea, *Anthrozoös* 5:4 (1992), pp245–253
tandfonline.com/doiabs/10.2752/089279392787011287?journalCode=rfan20

4.11 Why do dogs eat poo?

'Social organization of African Wild Dogs (*Lycaon pictus*) on the Serengeti Plains, Tanzania 1967–1978' by Lory Herbison Frame, James R Malcolm, George W Frame & Hugo Van Lawick, *Ethology* 50:3 (1979), pp225–249
onlinelibrary.wiley.com/doi/abs/10.1111/j.1439-0310.1979.tb01030.x

'Territoriality and scent marking behavior of African Wild Dogs in northern Botswana' by Margaret Parker, *Graduate Student Theses, Dissertations, & Professional Papers*, 954 (University of Montana, 2010)
scholarworks.umt.edu/cgi/viewcontent.cgi?article=1973&context=etd

5.01 Do dogs feel guilt?

'Disambiguating the "guilty look": Salient prompts to a familiar dog behaviour' by Alexandra Horowitz, *Behavioural Processes* 81:3 (2009), pp447–452
sciencedirect.com/science/article/abs/pii/S0376635709001004

'Behavioral assessment and owner perceptions of behaviors associated with guilt in dogs' by Julie Hecht, Ádám Miklósi & Márta Gács, *Applied Animal Behaviour Science* 139 (2012), pp134–142
etologia.elte.hu/file/publikaciok/2012/HechtMG2012.pdf

'Jealousy in Dogs' by Christine R Harris & Caroline Prouvost, *PLOS ONE* 9(7) (2014), e94597
journals.plos.org/plosone/article?id=10.1371/journal.pone.0094597

'Dogs understand fairness, get jealous, study finds'
npr.org/templates/story/story.php?storyId=97944783&t=1608741741655

'Shut up and pet me! Domestic dogs (*Canis lupus familiaris*) prefer petting to vocal praise in concurrent and single-alternative choice procedures' by Erica N Feuerbacher & Clive DL Wynn, *Behavioural Processes* 110 (2015), pp47–59
blog.wunschfutter.de/blog/wp-content/uploads/2015/02/Shut-up-and-pet-me.pdf

5.03 What does left- or right-handed tail wagging mean?

'Hemispheric Specialization in Dogs for Processing Different Acoustic Stimuli' by Marcello Siniscalchi, Angelo Quaranta & Lesley J Rogers, *PLOS ONE* 3(10) (2008), e3349
journals.plos.org/plosone/article?id=10.1371/journal.pone.0003349

'Lateralized Functions in the Dog Brain' by Marcello Siniscalchi, Serenella D'Ingeo & Angelo Quaranta, *Symmetry* 9(5) (2017), 71
mdpi.com/2073-8994/9/5/71/htm

5.04 How clever is your dog?

'Dogs recognize dog and human emotions' by Natalia Albuquerque *et al*, *Biology Letters* 12:1 (2016)
royalsocietypublishing.org/doi/10.1098/rsbl.2015.0883

'Female but not male dogs respond to a size constancy violation' by Corsin A Müller *et al*, *Biology Letters* 7:5 (2011)
royalsocietypublishing.org/doi/10.1098/rsbl.2011.0287

'Brain size predicts problem-solving ability in mammalian carnivores' by Sarah Benson-Amram *et al*, *Proceedings of the National Academy of Sciences of the United States of America* 113(9) (2016), 2532–2537
pnas.org/content/113/9/2532

'Free-ranging dogs are capable of utilizing complex human pointing cues' by Debottam Bhattacharjee *et al*, *Frontiers in Psychology* 10:2818 (2020)
frontiersin.org/articles/10.3389/fpsyg.2019.02818/full

5.05 Does your dog love you (or just need you)?

'Oxytocin-gaze positive loop and the coevolution of human-dog bonds' by Miho Nagasawa *et al*, *Science* 348:6232 (2015), pp333–336
science.sciencemag.org/content/348/6232/333

'The genomics of selection in dogs and the parallel evolution between dogs and humans' by Guo-dong Wang *et al*, *Nature Communications* 4, 1860 (2013)
nature.com/articles/ncomms2814

'Scent of the familiar: An fMRI study of canine brain responses to familiar and unfamiliar human and dog odors' by Gregory S Berns, Andrew M Brooks & Mark Spivak, *Behavioural Processes* 110 (2015), pp37–46
sciencedirect.com/science/article/pii/S0376635714000473

'Dogs recognize dog and human emotions' by Natalia Albuquerque *et al*, *Biology Letters* 12:1 (2016)
royalsocietypublishing.org/doi/10.1098/rsbl.2015.0883

'An exploratory study about the association between serum serotonin concentrations and canine-human social interactions in shelter dogs (*Canis familiaris*)' by Daniela Alberghina *et al*, *Journal of Veterinary Behavior* 18 (2017), pp96–101
sciencedirect.com/science/article/abs/pii/S1558787816301514

'Empathic-like responding by domestic dogs (*Canis familiaris*) to distress in humans: an exploratory study' by Deborah Custance & Jennifer Mayer, *Animal Cognition* 15 (2012), 851–859
academia.edu/1632457/Empathic_like_responding_by_domestic_dogs_Canis_familiaris_to_distress_in_humans_an_exploratory_study

5.06 What is my dog thinking?

'Third-party social evaluations of humans by monkeys and dogs' by James R Anderson *et al*, *Neuroscience & Biobehavioral Reviews* 82 (2017), pp95–109
sciencedirect.com/science/article/abs/pii/S0149763416303578

'Voice-sensitive regions in the dog and human brain are revealed by comparative fMRI' by Attila Andics *et al*, *Current Biology* 24:5 (2014), pp574–578
sciencedirect.com/science/article/pii/S0960982214001237?via%3Dihub

'Empathic-like responding by domestic dogs (Canis familiaris) to distress in humans: an exploratory study' by Deborah Custance & Jennifer Mayer, *Animal Cognition* 15 (2012), 851–859
academia.edu/1632457/Empathic_like_responding_by_domestic_dogs_Canis_familiaris_to_distress_in_humans_an_exploratory_study

'Dogs can discriminate emotional expressions of human faces' by Corsin A Müller, Kira Schmitt, Anjuli LA Barber & Ludwig Huber, *Current Biology* 25:5 (2015), pp601–605
sciencedirect.com/science/article/pii/S0960982214016935?via%3Dihub

5.07 Why do dogs yawn?

'Social modulation of contagious yawning in wolves' by Teresa Romero, Marie Ito, Atsuko Saito & Toshikazu Hasegawa, *PLOS One* 9(8) (2014), e105963
ncbi.nlm.nih.gov/pmc/articles/PMC4146576/

'Familiarity bias and physiological responses in contagious yawning by dogs support link to empathy' by Teresa Romero, Akitsugu Konno & Toshikazu Hasegawa, *PLOS One* 8(8) (2013), e71365
journals.plos.org/plosone/article?id=10.1371/journal.pone.0071365

'Dogs catch human yawns' by Ramiro M Joly-Mascheroni, Atsushi Senju & Alex J Shepherd, *Biology Letters* 4:5 (2008)
royalsocietypublishing.org/doi/10.1098/rsbl.2008.0333

'A test of the yawning contagion and emotional connectedness hypothesis in dogs, *Canis familiaris*' by Sean J O'Hara & Amy V Reeve, *Animal Behaviour* 81:1 (2011), pp335–40
sciencedirect.com/science/article/abs/pii/S0003347210004483

'Auditory contagious yawning in domestic dogs (*Canis familiaris*): first evidence for social modulation' by Karine Silva, Joana Bessa & Liliana de Sousa, *Animal Cognition* 15:4 (2012), pp721–4
pubmed.ncbi.nlm.nih.gov/22526686/

'Familiarity-connected or stress-based contagious yawning in domestic dogs (*Canis familiaris*)? Some additional data' by Karine Silva, Joana Bessa & Liliana de Sousa, *Animal Cognition* 16 (2013), pp1007–1009
link.springer.com/article/10.1007/s10071-013-0669-0

'Contagious yawning, social cognition, and arousal: An investigation of the processes underlying shelter dogs' responses to human yawns' by Alicia Phillips Buttner & Rosemary Strasser, *Animal Cognition* 17:1 (2014), pp95–104
pubmed.ncbi.nlm.nih.gov/23670215/

5.10 Do dogs dream? If so, what about?

'Baseline sleep-wake patterns in the pointer dog' by EA Lucas, EW Powell & OD Murphree, *Physiology & Behavior* 19(2) (1977), pp285–91
pubmed.ncbi.nlm.nih.gov/203958/

'Temporally structured replay of awake hippocampal ensemble activity during rapid eye movement sleep' by Kenway Louie & Matthew A Wilson, *Neuron* 29 (2001), pp145–156
cns.nyu.edu/~klouie/papers/LouieWilson01.pdf

5.13 Why do dogs love to play?

'Why do dogs play? Function and welfare implications of play in the domestic dog' by Rebecca Sommerville, Emily A O'Connor & Lucy Asher, *Applied Animal Behaviour Science* 197 (2017), pp1–8
sciencedirect.com/science/article/abs/pii/S0168159117302575

'Intrinsic ball retrieving in wolf puppies suggests standing ancestral variation for human-directed play behavior' by Christina Hansen Wheat & Hans Temrin, *iScience* 23:2 (2020), 100811
sciencedirect.com/science/article/pii/S2589004219305577?via%3Dihub

'Partner preferences and asymmetries in social play among domestic dog, *Canis lupus familiaris*, littermates' by Camille Ward, Erika B Bauer & Barbara B Smuts, *Animal Behaviour* 76:4 (2008), pp1187–1199
sciencedirect.com/science/article/pii/S0003347208002741?via%3Dihub#bib19

'Squirrel monkey play-fighting: making the case for a cognitive training function for play' by Maxeen Biben in *Animal Play: Evolutionary, Comparative, and Ecological Perspectives* by M Bekoff & JA Byers (Eds) (Cambridge University Press, 1998), pp161–182
psycnet.apa.org/record/1998-07899-008

The Genesis of Animal Play: Testing the Limits by GM Burghardt (MIT Press, 2005)
mitpress.mit.edu/books/genesis-animal-play

'Playful defensive responses in adult male rats depend on the status of the unfamiliar opponent' by LK Smith, S-LN Fantella & SM Pellis, *Aggressive Behavior* 25:2 (1999), pp141–152
onlinelibrary.wiley.com/doi/abs/10.1002/%28SICI%2910982337%281999%2925%3A2%3C141%3A%3AAID-AB6%3E3.0.CO%3B2-S

'Play fighting does not affect subsequent fighting success in wild meerkats' by Lynda L Sharpe, *Animal Behaviour* 69:5 (2005), pp1023–1029
sciencedirect.com/science/article/abs/pii/S0003347204004609

5.16 Why do dogs chase their tails?

'Environmental effects on compulsive tail chasing in dogs'
journals.plos.org/plosone/article?id=10.1371/journal.pone.0041684

6.01 Dog smell

'The science of sniffs: disease smelling dogs'
understandinganimalresearch.org.uk/news/research-medical-benefits/the-science-of-sniffs-disease-smelling-dogs/

6.02 Can dogs really diagnose disease?

'Olfactory detection of human bladder cancer by dogs: proof of principle study' by Carolyn M Willis *et al*, *BMJ* 329(7468):712 (2004)
ncbi.nlm.nih.gov/pmc/articles/PMC518893/

Medical Detection Dogs
https://www.medicaldetectiondogs.org.uk/

6.03 Dog sight

'What do dogs (*Canis familiaris*) see? A review of vision in dogs and implications for cognition research' by Sarah-Elizabeth Byosiere, Philippe A Chouinard, Tiffani J Howell & Pauleen C Bennett, *Psychonomic Bulletin & Review* 25 (2018), pp1798–1813
link.springer.com/article/10.3758/s13423-017-1404-7

7.01 Why do dogs bark?

'Barking dogs as an environmental problem' by CL Senn & JD Lewin, *Journal of the American Veterinary Medicine Association* 166(11) (1975), pp1065-1068.
europepmc.org/article/med/1133065

7.02 What does your dog hear when you talk?

'Dog-directed speech: why do we use it and do dogs pay attention to it?' by Tobey Ben-Aderet, Mario Gallego-Abenza, David Reby & Nicolas Mathevon, *Proceedings of the Royal Society* B 284:1846 (2017)
royalsocietypublishing.org/doi/10.1098/rspb.2016.2429

'Neural mechanisms for lexical processing in dogs' by Attila Andics *et al*, *Science* 10.1126/science.aaf3777 (2016)
pallier.org/lectures/Brain-imaging-methods-MBC-UPF-2017/papers-for-presentations/Andics%20et%20al.%20-%202016%20-%20Neural%20mechanisms%20for%20lexical%20processing%20in%20dogs.pdf

'Shut up and pet me! Domestic dogs (*Canis lupus familiaris*) prefer petting to vocal praise in concurrent and single-alternative choice procedures' by Erica N Feuerbacher & Clive DL Wynn, *Behavioural Processes* 110 (2015), pp47–59
blog.wunschfutter.de/blog/wp-content/uploads/2015/02/Shut-up-and-pet-me.pdf

8.01 Cat people vs dog people

'Personalities of Self-Identified "Dog People" and "Cat People"' by Samuel D Gosling, Carson J Sandy & Jeff Potter, *Anthrozoös* 23(3) (2010), pp213-222
researchgate.net/publication/233630429_Personalities_of_Self-Identified_Dog_People_and_Cat_People

'Cat People, Dog People' (Facebook Research)
research.fb.com/blog/2016/08/cat-people-dog-people/

'Owner perceived differences between mixed-breed and purebred dogs' by Borbála Turcsán, Ádám Miklósi & Enikő Kubinyi, *PLOS ONE* 12(2) (2017), e0172720.
journals.plos.org/plosone/article?id=10.1371/journal.pone.0172720

'The personality of "aggressive" and "non-aggressive" dog owners' by Deborah L Wells & Peter G Hepper, *Personality and Individual Differences* 53:6 (2012), pp770–773
sciencedirect.com/science/article/abs/pii/S0191886912002875?via%3Dihub

'Birds of a feather flock together? Perceived personality matching in owner–dog dyads' by Borbála Turcsán *et al*, *Applied Animal Behaviour Science* 140:3–4 (2012), pp154–160
sciencedirect.com/science/article/abs/pii/S0168159112001785?via%3Dihub

'Personality characteristics of dog and cat persons' by Rose M Perrine & Hannah L Osbourne, *Anthrozoös* 11:1 (1998), pp33–40
tandfonline.com/doi/abs/10.1080/08927936.1998.11425085

8.02 How much does a dog cost to run?

rover.com/blog/uk/cost-of-owning-a-dog/

8.04 Do owners look like their dogs?

'Do dogs resemble their owners?' by Michael M Roy & JS Christenfeld Nicholas, *Psychological Science* 15:5 (2004)
journals.sagepub.com/doi/abs/10.1111/j.0956-7976.2004.00684.x

'Self seeks like: many humans choose their dog pets following rules used for assortative mating' by Christina Payne & Klaus Jaffe, *Journal of Ethology* 23 (2005), pp15–18
link.springer.com/article/10.1007/s10164-004-0122-6

8.05 Are dogs good for your health?

'Dog ownership and the risk of cardiovascular disease and death – a nationwide cohort study' by Mwenya Mubanga *et al*, *Scientific Reports* 7 (2017), 15821
nature.com/articles/s41598-017-16118-6?utm_medium=affiliate&utm_source=commission_junction&utm_campaign=3_nsn6445_deeplink_PID100080543&utm_content=deeplink

'Dog ownership associated with longer life, especially among heart attack and stroke survivors'
newsroom.heart.org/news/dog-ownership-associated-with-longer-life-especially-among-heart-attack-and-stroke-survivors

'Why having a pet is good for your health'
health.harvard.edu/staying-healthy/why-having-a-pet-is-good-for-your-health

'Children reading to dogs: A systematic review of the literature' by Sophie Susannah Hall, Nancy R Gee & Daniel Simon Mills, *PLOS ONE* 11(2) (2016), e0149759
journals.plos.org/plosone/article?id=10.1371/journal.pone.0149759

'Dog ownership and cardiovascular health: Results from the Kardiovize 2030 project' by Andrea Maugeri *et al*, *Mayo Clinic Proceedings: Innovations, Quality & Outcomes* 3:3 (2019), pp268–275
sciencedirect.com/science/article/pii/S2542454819300888

'Benefits of dog ownership: Comparative study of equivalent samples' by Mónica Teresa González Ramírez & René Landero Hernández, *Journal of Veterinary Behavior* 9:6 (2014), pp311–315
pubag.nal.usda.gov/catalog/5337636

'Pet ownership and the risk of dying from lung cancer, findings from an 18 year follow-up of a US national cohort' by Atin Adhikari *et al*, *Environmental Research* 173 (2019), pp379–386
sciencedirect.com/science/article/abs/pii/S0013935119300416

'The relationship between dog ownership, psychopathological symptoms and health-benefitting factors in occupations at risk for traumatization' by Johanna Lass-Hennemann, Sarah K Schäfer, M Roxanne Sopp & Tanja Michael, *International Journal of Environmental Research and Public Health* 17(7): 2562 (2020)
ncbi.nlm.nih.gov/pmc/articles/PMC7178020/

'Why do dogs play? Function and welfare implications of play in the domestic dog' by Rebecca Sommerville, Emily A O'Connor & Lucy Asher, *Applied Animal Behaviour Science* 197 (2017), pp1–8
sciencedirect.com/science/article/abs/pii/S0168159117302575

'Physical activity benefits from taking your dog to the park' by Jenny Veitch, Hayley Christian, Alison Carver & Jo Salmon, *Landscape and Urban Planning* 185 (2019), pp173–179
sciencedirect.com/science/article/abs/pii/S0169204618312805

8.06 What does your dog do when you go out?

'Separation anxiety in dogs'
rspca.org.uk/adviceandwelfare/pets/dogs/behaviour/separationrelatedbehaviour

8.07 The climate impact of your dog

'Environmental impacts of food consumption by dogs and cats' by Gregory S Okin, *PLOS ONE* 12(8) (2017), e0181301
journals.plos.org/plosone/article?id=10.1371/journal.pone.0181301

'The ecological paw print of companion dogs and cats' by Pim Martens, Bingtao Su & Samantha Deblomme, *BioScience* 69:6 (2019), pp467–474
academic.oup.com/bioscience/article/69/6/467/5486563

'The climate mitigation gap: education and government recommendations miss the most effective individual actions' by Seth Wynes & Kimberly A Nicholas, *Environmental Research Letters* 12:7 (2017)
iopscience.iop.org/article/10.1088/1748-9326/aa7541

9.02 Dogs vs Cats: social and medical

'Pet Population 2020' (PFMA)
pfma.org.uk/pet-population-2020

'Pet Industry Market Size & Ownership Statistics' (American Pet Products Association)
americanpetproducts.org/press_industrytrends.asp

9.03 Dogs vs Cats: physical head-to-head

'Dogs have the most neurons, though not the largest brain: trade-off between body mass and number of neurons in the cerebral cortex of large carnivoran species' by Débora Jardim-Messeder *et al*, *Frontiers in Neuroanatomy* 11:118 (2017)
frontiersin.org/articles/10.3389/fnana.2017.00118/full

'The role of clade competition in the diversification of North American canids' by Daniele Silvestro, Alexandre Antonelli, Nicolas Salamin & Tiago B Quental, Proceedings of the *National Academy of Sciences of the United States of America* 112(28) (2015), 8684-8689
pnas.org/content/112/28/8684

10.03 Why are dogs so greedy?

'A deletion in the canine POMC gene is associated with weight and appetite in obesity-prone labrador retriever dogs' by Eleanor Raffan *et al*, *Cell Metabolism* 23:5 (2016), pp893–900
cell.com/cell-metabolism/fulltext/S1550-4131(16)30163-2

10.04 What's in dog food?

'Identification of meat species in pet foods using a real-time polymerase chain reaction (PCR) assay' by Tara A Okumaa & Rosalee S Hellberg, *Food Control* 50 (2015), pp9–17
sciencedirect.com/science/article/abs/pii/S0956713514004666

'Animal by-products' (EU)
ec.europa.eu/food/safety/animal-by-products_en

'Pet food' (Food Standards Agency)
food.gov.uk/business-guidance/pet-food

Acknowledgements

This book is based on the work of thousands of wonderful researchers and authors who have put their expertise down in print, and although I've cited the main papers and books I've referenced, there were hundreds more that were vital for understanding this fascinating world. It's both odd and sad that most of this is publicly-funded research, yet scientific publishers make huge profits from it and effectively ring-fence that knowledge from the public. Let's hope this changes sooner rather than later.

Thanks so much to the wonderful Sarah Lavelle, Stacey Cleworth and Claire Rochford at Quadrille for such enthusiasm for my weird fascinations and for putting up with both me and my inability to take deadlines seriously. And to Luke Bird for taking on another weird project with such good grace.

Thanks so much to my gorgeous girls Daisy, Poppy and Georgia, for leaving me alone at the end of the garden to write, and for enduring the breathlessly enthusiastic stream of facts I subjected them to over dinner. Thanks also to Blue and Cheeky for enduring my constant poking whilst testing vomeronasal organs, nictitating membranes, fur-counts, cross-species communication and claw-retracability. Thanks also to Brodie Thomson, Eliza Hazlewood and Coco Ettinghausen and, as always, to the amazing and wonderfully supportive crew at DML: Jan Croxson, Borra Garson, Lou Leftwich and Megan Page.

Lastly, thanks so much to the brilliant audiences who've come along to my shows and laughed their pants off whilst we've explored some utterly fascinating or revolting science live on stage. I love you people.

Index

Publishing Director: Sarah Lavelle
Head of Design: Claire Rochford
Designer and Illustrator: Luke Bird
Copy Editor: Nick Funnell
Editor: Stacey Cleworth
Editorial Assistant: Sofie Shearman
Head of Production: Stephen Lang
Production Controller: Katie Jarvis

First published in 2021 by Quadrille, an imprint of Hardie Grant Publishing

Quadrille
52–54 Southwark Street
London SE1 1UN
quadrille.com

Cataloguing in Publication Data: a catalogue record for this book is available from the British Library.

ISBN: 978 1 78713 633 5
Printed in China